Failure Analysis of Biometals

Failure Analysis of Biometals

Special Issue Editor

Reza Hashemi

MDPI • Basel • Beijing • Wuhan • Barcelona • Belgrade

Special Issue Editor
Reza Hashemi
College of Science and
Engineering, Medical Device
Research Institute,
Flinders University
Australia

Editorial Office
MDPI
St. Alban-Anlage 66
4052 Basel, Switzerland

This is a reprint of articles from the Special Issue published online in the open access journal *Metals* (ISSN 2075-4701) from 2013 to 2014 (available at: https://www.mdpi.com/journal/metals/special_issues/failure_biometals).

For citation purposes, cite each article independently as indicated on the article page online and as indicated below:

LastName, A.A.; LastName, B.B.; LastName, C.C. Article Title. *Journal Name* **Year**, *Article Number*, Page Range.

ISBN 978-3-03936-499-2 (Pbk)
ISBN 978-3-03936-500-5 (PDF)

Cover image courtesy of Reza Hashemi.

Contents

About the Special Issue Editor

Reza Hashemi (former name: Hashemi Oskouei) is an academic at the College of Science and Engineering and Research Leader in the Medical Device Research Institute at Flinders University, Adelaide, Australia. He holds a PhD in Mechanical Engineering from Monash University in Australia. His primary research interests include the mechanical behaviour of metallic biomaterials as well as their processing, characterisation, testing and failure, fretting wear and corrosion in metallic implants, and fatigue and fracture analysis. Dr Hashemi's research is both experimental and computational (finite element analysis and modelling), and he has made significant contributions to the field of materials characterisation and failure, in addition to finite element simulations of complex phenomena such as fretting wear and fretting corrosion.

metals

MDPI

Editorial

Failure Analysis of Biometals

Reza Hashemi

College of Science and Engineering, Medical Device Research Institute, Flinders University, Tonsley SA 5042, Australia; reza.hashemi@flinders.edu.au; Tel.: +61-8-82012782

Received: 30 April 2020; Accepted: 15 May 2020; Published: 19 May 2020

1. Introduction and Scope

Metallic biomaterials (biometals) are widely used for the manufacture of medical implants, ranging from load-bearing orthopaedic prostheses to dental and cardiovascular implants, because of their favourable combination of properties including high strength, fracture toughness, biocompatibility, and wear and corrosion resistance. Additionally, they can be fabricated using well-established techniques (such as casting and forging), and recently, additive manufacturing to produce complex and customised implants. Examples of metals and metal alloys that are used for the fabrication of implants include the following: Ti-based alloys (e.g., Ti6Al4V and Ti6Al7Nb), Co-based alloys (e.g., CoCrMo and CoNi), austenitic stainless steels (e.g., SS316L), Zr-Nb alloys, Ni-Ti Alloys, Mg alloys, porous tantalum foams, and precious metals and alloys.

Owing to the significant consequences of implant material failure/degradation, in terms of both personal and financial burden, failure analysis of biometals (in-vivo, in-vitro, and retrieval) has always been of paramount importance in order to understand the failure mechanisms and implement suitable solutions with the aim to improve the longevity of implants in the body. This Special Issue presents some of the latest developments and findings in the area of biometals failure including fatigue, fracture, corrosion, and fretting wear on a range of conventional biometals as well as porous materials and new generation titanium alloys.

2. Contributions

The Special Issue "Failure Analysis of Biometals" comprises ten original research articles [1–10] covering a great common range of metallic biomaterials (Ti alloys, CoCrMo alloys, Mg alloys, NiTi alloys) and their failure mechanisms (corrosion, fatigue, fracture, and fretting wear) that commonly occur in medical implants and surgical instruments. This collection of studies also includes two review papers [9,10]: the corrosion behaviour of new generation low modulus titanium alloys for implant applications, and the three-dimensional (3D) printed acetabular cups for hip replacement implants reviewing the clinical use of 3D printing in orthopaedics.

Starting with research articles, Alqedairi et al. [1] studied the cyclic fatigue behaviour of endodontic rotary files made of nickel titanium (NiTi) alloys. Owing to the cyclic rotation of this dental instrument within the curved canal, fatigue fracture can occur over time. The failure can be either flexural or torsional in its loading nature. Three different rotary instruments including ProTaper Universal (PTU), ProTaper Gold (PTG), and ProTaper Next (PTN) were assessed in this work. Artificial canals were machined in stainless steel blocks. Fifteen rotary instruments of each type (five types of PTU and PTG and three types of PTN) were rotated with 300 rpm until fracture. Weibull reliability analysis and the probabilities of survival calculated for the PTU, PTG, and PTN instruments showed the PTG series to offer a higher reliability than the PTU and PTN series. The PTG instruments were found to have a superior cyclic fatigue behaviour when compared with the PTU and PTN series. The higher fatigue resistance of the PTG and PTN instruments was attributed to the thermomechanical treatment performed on the NiTi alloy of these instruments. Fatigue is understood to be the most common failure mechanism in rotary instruments; therefore, international standards for testing against fatigue failure of these medical devices should be established in order to reduce considerable differences in their behaviour.

In a two-part study conducted by Goswami and co-authors [2,3], the failure of a stainless steel 316 L proximal humerus internal locking system (PHILOS) plate and screws, which had been used for a pantalar arthrodesis, was investigated. The research employed both experimental (SEM/EDS, microstructural analysis using electron backscatter diffraction (EBSD), hardness tests, and fractography) and computational (finite element modelling) approaches. The results of fractography, particularly SEM investigations, indicated the occurrence of corrosion fatigue failure initiated by crack initiations in the distal region of the steel plate, leading to crack propagations towards the proximal region and a final brittle fracture. Crack initiations in the plate were reported to be the result of the inclusions and corrosion pits. The failure of the screws was because of overloading, which occurred ahead of the plate from the proximal end. Finite element analysis (FEA) on the implant system, implemented using ANSYS, showed increased von Mises stresses in the cortical screws as the angle between the screws and the plates increased. Moreover, the stress magnitudes were found to be lower (by 25.5%) in the locking screws when compared with the cortical screws, which was understood to be because of the fixed angles of the locking screws onto the plate (less range of motion).

Returning the focus again to nickel titanium alloys in this Special Issue, the in vitro corrosion assessment of porous NiTi structures was studied by Ibrahim et al. [4] for bone fixation applications. The structures were fabricated by additive manufacturing, which is expanding very rapidly nowadays. It is, however, noted that additively manufactured NiTi structures still have some issues such as poor surface quality and presence of impurities and defects. Employing the selective laser melting (SLM) technique, NiTi samples in both the bulk and porous forms (porosity levels of 15–50%) were fabricated. The electrochemical corrosion characteristics of these SLM NiTi samples were found to be similar to those of conventionally fabricated NiTi samples. The 50% porous structures showed the highest Ni ion release level owing to their biggest surface area exposed to the corrosive environment. The main finding in this work was that the SLM manufacturing process employed to fabricate NiTi structures for bone fixation applications did not cause a significant deterioration in their corrosion resistance.

In medical implants, where there is contact between the implant material and living bone, a stable implant–bone interface is essential for clinical success of fixation by osseointegration and bone ingrowth. A second important factor vital for achieving success is that there be a minimal mismatch between the mechanical properties of the implanted prostheses and the host skeleton. If there is a substantial discrepancy between the said properties, significant stress shielding can occur, causing adverse effects on the implant and/or the host skeleton. Additive manufacturing can be wisely employed to fabricate stiffness-modulated implants to minimise stress shielding failure, which is one of the interesting areas of research at present. In a study by Jahadakbar et al. [5], bone fixation plates were designed and manufactured out of NiTi alloys using additive manufacturing (SLM method), such that the stiffness of the Nitinol plates was modulated. Five different porosity levels (17%, 20%, 24%, 27%, and 30%) as well as a bulk plate (0% porosity) were designed and analysed using finite element modelling (Abaqus software), showing a good agreement with the experimental results of mechanical testing. Following the model verification, Ni-rich fixation plates were manufactured, offering a superelastic behaviour. Differential scanning calorimetry (DSC) was employed in order to identify the transformation temperatures (TTs) from −90 to 100 °C in a nitrogen atmosphere. The DSC results showed a small variation in the transformation temperature of different points in the fixation plates owing to the various thermal histories that the complex plates experienced during the additive manufacturing process. A post-processing heat treatment may thus be required in order to achieve homogeneity in the as-fabricated parts.

Magnesium (Mg)-based alloys exhibit biodegradable and biocompatible characteristics, enabling them to be used in degradable implants that can dissolve in the body after the treatment of a medical condition. However, there are still a number of aspects that need to be further researched and addressed (e.g., mechanical and corrosion properties and degradation behaviour). Taking into account that calcium (Ca) is an important element in the bone structure, Chen et al. [6] developed a magnesium-based alloy (Mg-1Ca-0.5Zr) using casting. Zirconium (Zr) was also added to enhance the grain refinement in the alloy. The main aim was to assess the mechanical properties and surface corrosion mechanism in the developed alloy. A heat treatment (400 °C for 8 h, followed by quenching in water) was applied to the alloy to improve its mechanical properties. In addition to tensile tests, erosion wear tests were performed on the alloy samples with and without the heat treatment. Moreover, a potentiodynamic polarization test was performed on these samples and a pure Mg sample. The heat treatment was found to enhance the ductility and reduce the corrosion rate of the alloy. Moreover, it was reported that the Mg alloy subjected to the heat treatment formed a protective calcium phosphate film when immersed and tested in a simulated body fluid. This protective layer decreased the corrosion rate considerably.

In an investigation on retrieved total hip replacement implants, particularly the metallic taper junction known as the head–neck junction, corrosion damage to the neck part of 137 femoral stem implants was analysed using the Goldberg's scoring method [7]. The studied stems were made of three biometals including CoCrMo, stainless steel (SS), and titanium alloy. The neck surface was divided into eight distinct zones to statistically study the distribution and severity of corrosion damage. The distal region was found to have more corrosion damage compared with the proximal region of the neck. The most severe corrosion also occurred in the medial distal zone. This study suggests that retrieval studies of head–neck taper junctions should assess the corrosion damage in various zones of metallic implants separately.

In the same area of the head–neck taper junction in hip implants, Fallahnezhad et al. [8] studied material loss as a result of fretting wear with a focus on the role of assembly force (impaction force applied by surgeons to assemble the junction intraoperatively). Both the head and neck components were made out of CoCrMo alloy, with an angular mismatch of 0.01°. Developing an adaptive finite element model for fretting wear, four assembly forces (2, 3, 4, and 5 kN) were applied to the junction followed by a walking gait loading (1,025,000 cycles). The results showed the direct effect of assembly force; the higher the force, the greater the material removal owing to fretting wear. It was discussed that a high assembly force can generally improve the stability of the junction; however, it may also enhance the wear damage to the material. This study did not include corrosion in the simulations; thus, further research is suggested to create novel models to capture both fretting wear and corrosion simultaneously.

In a review on the clinical use of 3D printing (additive manufacturing) in orthopaedics, Dall'Ava et el. [9] focused on titanium acetabular cup implants used in total hip replacement, where they compared 3D printing with conventional manufacturing. This review defined the rationale of additively manufactured acetabular cups from both the clinical and engineering perspectives. A number of interesting aspects associated with the topic were discussed and summarised. These include the key manufacturing-related variables that can have an influence on the characteristics and properties of the fabricated implants, and the limitations associated with this manufacturing technology. Additively manufactured titanium cups have presented promising early clinical outcomes. It is, however, important to note that more detailed studies are still needed to look at their failure and long-term performance in the body.

Finally, this Special Issue presents a review article [10] on the corrosion behaviour of new generation titanium alloys (β-type offering low Young's modulus) that can be desirably used for medical implants.

There are some existing concerns about the toxicity of the two alloying elements (aluminium and vanadium) in the most commonly used titanium alloy in medical applications (Ti-6Al-4V). Furthermore, the stiffness (Young's modulus) of this alloy, which is approximately 110 GPa, is much higher than the typical stiffness of the bone (10–30 GPa). This can result in stress shielding under the loads of physical activities and, consequently, prosthesis loosening, bone loss, and fracture failure. To address these issues, extensive research has been done to develop new generation β-type titanium alloys with lower levels of stiffness. These alloys contain beta-stabilising elements, for example, Nb, Ta, and Zr, which are also non-allergic and non-toxic. Although there has been a lot of work around the improvement of mechanical properties in these alloys, their corrosion behaviour still needs further research (given that the implant working environment is corrosive). In this article, Afzali et al. [10] reviewed and discussed a number of key factors (fabrication process, chemical composition, passive layer, mechanical treatments, body electrolyte properties, and constituent phases) influencing the corrosion behaviour/resistance of new generation titanium alloys. The effects of α and β phases and their dissolution rates on the oxide layer and corrosion behaviour were also reviewed. It was recommended that the microstructure of these new generation alloys should contain suitable amounts of α and β phases to achieve a high corrosion resistance, as well as a stable oxide layer.

3. Conclusions

Failure Analysis of Biometals presents a collection of studies covering a wide range of failure mechanisms in medical implant materials. The contributions reflect the profound interest in this area aiming to address current issues in biometals, manufacturing techniques, and implant applications, while employing various research methodologies. Challenges still remain, however, there are also great opportunities for research to better analyse and understand failures in biometals and come up with successful engineering solutions.

Acknowledgments: As Guest Editor, I would like to thank all the researchers who contributed their work to this Special Issue; and also, the reviewers who provided feedback to improve the quality of the articles. I would also like to thank the MDPI *Metals* Editorial Team, especially the Managing Editor Mrs. Sunny He, for their great management and support in the publication process.

Conflicts of Interest: The author declares no conflict of interest.

References

1. Alqedairi, A.; Alfawaz, H.; Bin Rabba, A.; Almutairi, A.; Alnafaiy, S.; Khan Mohammed, M. Failure Analysis and Reliability of Ni–Ti-Based Dental Rotary Files Subjected to Cyclic Fatigue. *Metals* **2018**, *8*, 36. [CrossRef]
2. Ina, J.; Vallentyne, M.; Hamandi, F.; Shugart, K.; Boin, M.; Laughlin, R.; Goswami, T. Failure Analysis of PHILOS Plate Construct Used for Pantalar Arthrodesis Paper I—Analysis of the Plate. *Metals* **2018**, *8*, 180. [CrossRef]
3. Hamandi, F.; Laughlin, R.; Goswami, T. Failure Analysis of PHILOS Plate Construct Used for Pantalar Arthrodesis Paper II—Screws and FEM Simulations. *Metals* **2018**, *8*, 279. [CrossRef]
4. Ibrahim, H.; Jahadakbar, A.; Dehghan, A.; Moghaddam, N.S.; Amerinatanzi, A.; Elahinia, M. In Vitro Corrosion Assessment of Additively Manufactured Porous NiTi Structures for Bone Fixation Applications. *Metals* **2018**, *8*, 164. [CrossRef]
5. Jahadakbar, A.; Nematollahi, M.; Safaei, K.; Bayati, P.; Giri, G.; Dabbaghi, H.; Dean, D.; Elahinia, M. Design, Modeling, Additive Manufacturing, and Polishing of Stiffness-Modulated Porous Nitinol Bone Fixation Plates Followed by Thermomechanical and Composition Analysis. *Metals* **2020**, *10*, 151. [CrossRef]
6. Chen, Y.-T.; Hung, F.-Y.; Syu, J.-C. Biodegradable Implantation Material: Mechanical Properties and Surface Corrosion Mechanism of Mg-1Ca-0.5Zr Alloy. *Metals* **2019**, *9*, 857. [CrossRef]
7. Milimonfared, R.; Oskouei, R.H.; Taylor, M.; Solomon, L.B. The Distribution and Severity of Corrosion Damage at Eight Distinct Zones of Metallic Femoral Stem Implants. *Metals* **2018**, *8*, 840. [CrossRef]
8. Fallahnezhad, K.; Oskouei, R.H.; Badnava, H.; Taylor, M. The Influence of Assembly Force on the Material Loss at the Metallic Head-Neck Junction of Hip Implants Subjected to Cyclic Fretting Wear. *Metals* **2019**, *9*, 422. [CrossRef]
9. Dall'Ava, L.; Hothi, H.; Di Laura, A.; Henckel, J.; Hart, A. 3D Printed Acetabular Cups for Total Hip Arthroplasty: A Review Article. *Metals* **2019**, *9*, 729. [CrossRef]
10. Afzali, P.; Ghomashchi, R.; Oskouei, R.H. On the Corrosion Behaviour of Low Modulus Titanium Alloys for Medical Implant Applications: A Review. *Metals* **2019**, *9*, 878. [CrossRef]

metals

MDPI

Article

Failure Analysis and Reliability of Ni–Ti-Based Dental Rotary Files Subjected to Cyclic Fatigue

Abdullah Alqedairi [1,*], Hussam Alfawaz [1], Amani bin Rabba [1], Areej Almutairi [1], Sarah Alnafaiy [1] and Muneer Khan Mohammed [2]

1 Department of Restorative Dental Sciences, College of Dentistry, King Saud University, P.O. Box 60169, Riyadh 11545, Saudi Arabia; halfawaz1@ksu.edu.sa (H.A.); dr.amooon@hotmail.com (A.b.R.); tootah1410@hotmail.com (A.A.); saranafaiy@yahoo.com (S.A.)

2 Princess Fatima Alnijiris's Research Chair for Advanced Manufacturing Technology, Advanced Manufacturing Institute, King Saud University, P.O. Box 800, Riyadh 11421, Saudi Arabia; muneer0649@gmail.com

* Correspondence: aalqedairi@ksu.edu.sa; Tel.: +966-114-677-420

Received: 12 November 2017; Accepted: 3 January 2018; Published: 6 January 2018

Abstract: The cyclic fatigue resistance of ProTaper Universal (PTU), ProTaper Gold (PTG), and ProTaper Next (PTN) nickel titanium (NiTi) rotary files was evaluated. Fifteen instruments of each type were selected, totaling 195 files. The instruments were rotated until fracture in an artificial canal with dimensions corresponding to the dimensions of each instrument tested: +0.1 mm in width and 0.2 mm in depth, an angle of curvature of 45°, a radius of curvature of 5 mm, and a center of curvature 5 mm from the instrument tip. The fracture surfaces of three representative samples of each subgroup were examined using scanning electron microscopy (SEM). Time to fracture was analyzed via analysis of variance and Tukey's tests ($P < 0.05$). PTG F1 and F2 had significantly higher resistance than PTU F1 and PTN X2, and PTU F2 and PTN X3, respectively. PTN X2 showed a significantly higher resistance than PTU F1. The PTG series demonstrated superior cyclic fatigue (CF) behavior compared with that of the PTU and PTN series.

Keywords: cyclic; fatigue; gold; next; ProTaper; universal

1. Introduction

Due to iatrogenic procedural errors associated with the material stiffness of stainless-steel instruments, nickel–titanium (NiTi) material was introduced in the production of endodontic files [1]. The main characteristics of NiTi rotary instruments include memory shape, superior elasticity, and a centered canal preparation. In particular, the elastic flexibility of NiTi instruments is two to three times higher than that of stainless-steel instruments due to their lower modulus of elasticity [2,3]. The material properties of NiTi and stainlessness rotary files are presented in Table 1 [4].

Despite the elastic flexibility of NiTi rotary systems, instrument fracture has been reported [5,6]. The failure of rotary NiTi files can be either flexural (cyclic) or torsional [7]. The majority of studies have shown that cyclic fatigue (CF) fracture occurs when an instrument is flexed in the maximum curvature region of the canal while rotating freely, resulting in repeated tension–compression cycles [6,8]. The tension occurs on the part of the instrument on the outside of the curve, whereas the compression occurs on the other part on the inside of the curve. Therefore, these repeated cycles, caused by the rotation of the instrument within the curved canal, result in instrument fracture due to the increase in the cyclic fatigue of the instrument over time. Torsional fatigue occurs when an instrument tip is locked in a canal, while the body continues to rotate. Therefore, fracture of the tip becomes unpreventable when the torque exerted by the handpiece exceeds the elastic limit of the metal [8]. However, one of the limitations in in vitro studies of the cyclic fatigue behavior of NiTi instruments is the difficulty of assessing the clinical relevance of published tests results, where there are several factors, including torsional fatigue, at play at the same time.

Table 1. Properties of NiTi and stainless steel rotary files.

Properties	Ni–Ti Alloy	Stainless Steel
Ultimate tensile strength	~1240 MPa	~760 MPa
Density	6.45 gm/cm^3	8.03 gm/cm^3
Recoverable elongation	8%	0.8%
Effective modulus	~48 GPa	~193 GPa
Coefficient of thermal expansion	$6.6 \times 10^{-6} - 11 \times 10^{-6}$ °C^{-1}	17.3×10^{-6} °C^{-1}
Micro-hardness	303–362 VHN	522–542 VHN

The mechanical behavior and elastic flexibility of the NiTi alloy were improved by changing the transformation behavior of the alloy through heat treatment [9]. The NiTi alloy contains three microstructural phases (austenite, martensite, and R phase). Instruments in the martensite phase can be soft, ductile, and easily deformed and can recover their shape upon heating above the transformation temperature. Compared with conventional super-elastic NiTi, which shows a finish temperature of 16–31 °C [7,9], controlled memory wire and M-wire instruments show increased austenite transformation finish temperatures of approximately 55 and 50 °C, respectively [10]. Therefore, at body temperature, the conventional super-elastic NiTi file has an austenite structure, whereas an NiTi file with thermal processing is essentially in the martensite phase [9].

ProTaper Universal (PTU) and ProTaper Gold (PTG) rotary instruments possess the same geometries; however, PTG instruments have been metallurgically enhanced through heat-treatment technology in an attempt to improve flexibility, resistance to CF, and durability [7,11]. ProTaper Next (PTN) instruments are made of M-wire, which is fabricated by the thermomechanical processing of NiTi wire blanks [5]. In addition, fracture resistance has been improved in PTN instruments due to the unique asymmetrical rotary motion and reduced contact points between the instrument and root canal walls [5].

In the endodontic literature, rotational bending is applied to test for CF in NiTi rotary instruments. Several devices and methods have been used to evaluate the in vitro CF fracture resistance of NiTi rotary endodontic instruments [8]. In addition to two important parameters used to determine the shape of the root canal, i.e., the angle and radius of curvature [6], some studies have reported that the results obtained might be unreliable and inconsistent if the established device parameters do not follow each instrument's morphologic and geometric features [8]. To overcome this problem, multiple devices with artificial canals that have dimensions that exceed those of the tested instruments by 0.1–0.3 mm have been used [12–14].

No previous study has compared the CF resistance of all the ProTaper instruments among the three different generations. Therefore, the aim of the present study was to assess the CF behavior of the PTU, PTG, and PTN NiTi rotary files.

2. Materials and Methods

2.1. *Preparation of Artificial Canals*

The laser micromachining technique was used to machine artificial canals in stainless-steel plates with dimensions of 100 mm × 50 mm × 10 mm. Machining was performed using the LASERTEC 40 (Deckel Maho Gildemeister, Hamburg, Germany), which consists of a Q-switched Neodymium-doped Yttrium Aluminum Garnt (Nd: Y3Al5O12 (Nd: YAG)) laser operating at a wavelength of 1064 nm with a maximum average power of 30 W.

The artificial canal to be machined was modeled using CATIA V5® software (Dassault Systèmes, Version 5, Vélizy, France), and laser path programming was performed with a Standard Triangle

Language file of the proprietary machine software. After the laser process parameters were established, the laser was focused on the workpiece with the aid of a galvano scanner, and the canal was then machined layer by layer [15].

The artificial canals were machined in stainless-steel blocks with dimensions corresponding to the dimensions of the instrument tested: +0.1 mm in width and +0.2 mm in depth, with an angle of curvature of 45°, a radius of curvature of 5 mm, and a center of curvature 5 mm from the tip of the instrument [6,8] (Figure 1).

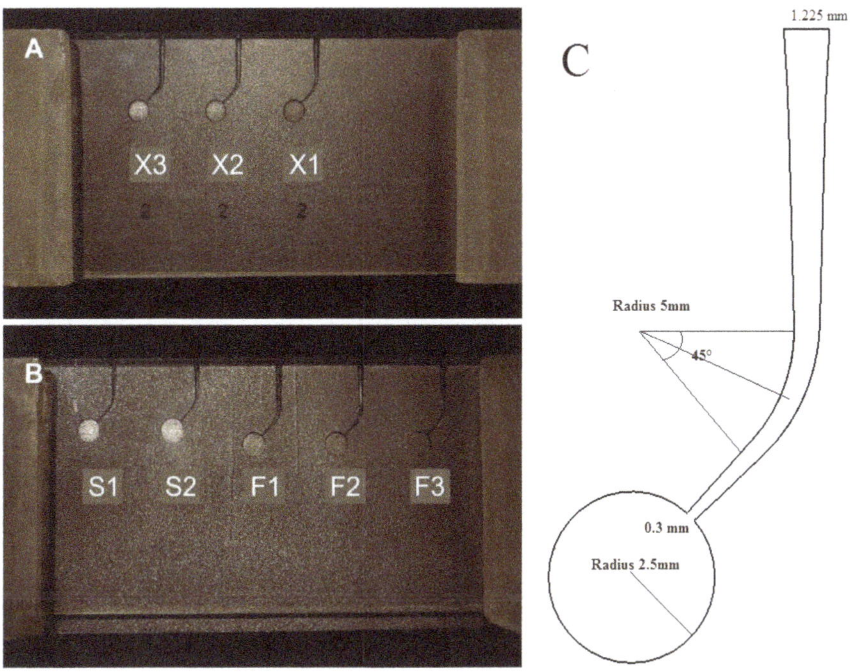

Figure 1. Custom-made stainless-steel blocks with dimensions corresponding to the dimensions of ProTaper Next (PTN) (**A**), ProTaper Gold (PTG), and ProTaper Universal (PTU) (**B**): +0.1 mm in width and +0.2 mm in depth, with an angle of curvature of 45°, a radius of curvature of 5 mm, and a center of curvature 5 mm from the tip of the instrument. (**C**) Two-dimensional draft of artificial canal for PTU F1 instrument.

The dimensions of the PTU and PTG instruments were recorded according to the manufacturer as shown in Table 2. The actual dimension for the PTN from the manufacturer along with the maximum diameters of the PTN instruments measured using Digimizer® software (MedCalc Software, version 4.5., Ostend, Belgium) are shown in Table 3.

Table 2. Dimensions of PTU and PTG from the manufacturer.

Active Part Length (mm)	Diameter (mm)				
	S1	S2	F1	F2	F3
0	0.170	0.200	0.200	0.250	0.300
1	0.190	0.240	0.270	0.330	0.390
2	0.220	0.285	0.340	0.410	0.480
3	0.260	0.335	0.410	0.490	0.570
4	0.305	0.390	0.465	0.550	0.640
5	0.355	0.450	0.520	0.610	0.710
6	0.415	0.510	0.575	0.665	0.760
7	0.485	0.570	0.630	0.720	0.810
8	0.565	0.630	0.685	0.775	0.860
9	0.655	0.690	0.740	0.830	0.910
10	0.755	0.760	0.795	0.885	0.960
11	0.855	0.850	0.850	0.940	1.010
12	0.960	0.955	0.905	0.995	1.060
13	1.075	1.070	0.960	1.050	1.110
14	1.185	1.185	1.015	1.105	1.160
15			1.070	1.160	1.210
16			1.125	1.215	1.260

Table 3. Dimensions of the PTN from the manufacturer.

Active Part Length (mm)	Diameter (mm)					
	X1		X2		X3	
	Actual	Maximum	Actual	Maximum	Actual	Maximum
16	1.16	1.26	1.2	1.3	1.2	1.34
13	0.98	1.06	1.11	1.15	1.09	1.14
9	0.7	0.76	0.84	1.06	0.89	1
6	0.49	0.534	0.63	0.7	0.71	0.78
3	0.31	0.35	0.43	0.45	0.53	0.65
1	0.21	0.23	0.31	0.34	0.38	0.52
0	0.17	0.17	0.25	0.25	0.3	0.3

2.2. *Cyclic Fatigue Testing*

Fifteen rotary instruments of each type (PTU S1, S2, F1, F2, and F3, PTG S1, S2, F1, F2 and F3, and PTN X1, X2, and X3), totaling 195 instruments of 25 mm in length, were used in this study.

Stainless-steel blocks were attached to a main frame to which a mobile support for the handpiece was connected. The dental handpiece was mounted on a mobile device that allowed for the simple placement of each instrument inside the artificial canal as shown in Figure 2. To prevent the instruments from slipping out and to allow for observation of the instruments, the artificial canals were covered with glass.

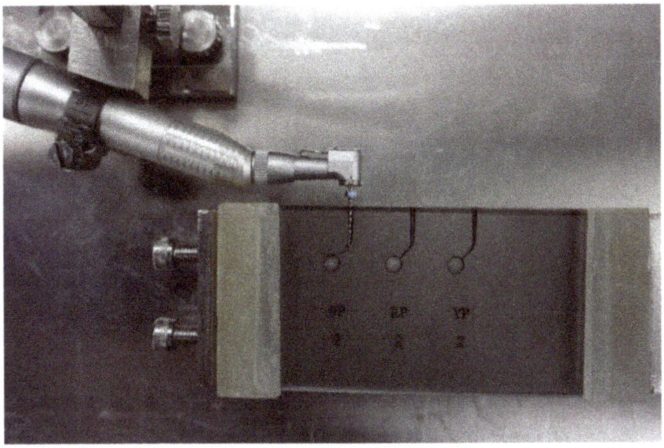

Figure 2. CF testing device illustrating positioning of dental handpiece, NiTi rotary instrument, and stainless steel block.

A pilot study was conducted to confirm the reliability of the CF device. All of the instruments were rotated at the speed recommended by the manufacturer (300 rpm) until fracture. The artificial canals were lubricated with synthetic oil (3-In-One Multi-Purpose Oil, WD-40®, San Diego, CA, USA) to reduce the friction of the tested file against the artificial canal walls. The motor and timer were then simultaneously activated. During each test, the instrument was monitored and visualized through the glass until fracture occurred, and the time to fracture was registered in seconds. Figure 3 shows the rotary files before and after fracture. The fractured surface was examined using SEM (JEOL 6360LV Scanning Electron Microscope, Tokyo, Japan) after preparation with ≥99.8% ethanol.

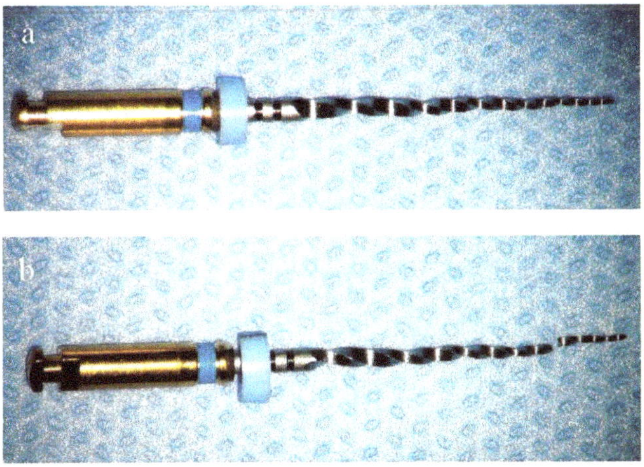

Figure 3. ProTaper Next X3 before (**a**) and after (**b**) fracture.

Statistical analysis of the empirical data is essential for the proper interpretation and prediction of results. There are many statistical methods such as analysis of variance (ANOVA), regression analysis, and correlation for analyzing data and representation of results [16]. Reports are available on the use of statistical methods for cyclic fatigue failure analysis [17] and fatigue life prediction [18].

In this work, one-way ANOVA and Tukey's tests were performed to analyze and compare the means. Statistical significance was set at $P < 0.05$. Weibull reliability analysis was performed and the probability of survival was calculated for the tested instruments.

3. Results

The mean times to fracture and standard deviations for the PTU, PTG, and PTN instruments are presented in Table 4. The CF behaviors of the PTU, PTG, and PTN series are presented in Figure 4. Comparing the instruments with similar D5 ±0.01 mm, one-way ANOVA and Tukey's post-hoc tests showed that PTG F1 and F2 had significantly higher CF resistance than PTU F1 and PTN X2, and PTU F2 and PTN X3, respectively. PTN X2 showed a significantly higher CF resistance than PTU F1. However, there was no significant difference between PTU F2 and PTN X3 in terms of CF resistance.

Table 4. Instrument type, sample size, time to fracture (seconds; mean ± SD), and Weibull calculations.

Instrument	N	Mean ± SD	Weibull Modulus	R-Squared	Predicted Time in Seconds for 99% Survival
PTU					
S1	15	166.07 ± 34.3	4.809	0.914	69
S2	15	170.40 ± 21.9	8.500	0.924	104
F1	15	101.47 ± 13.6	8.528	0.986	62
F2	15	93.20 ± 15.2	6.405	0.965	48
F3	15	87.20 ± 13.8	6.338	0.918	44
PTG					
S1	15	352.5 ± 57.4	6.357	0.916	181
S2	15	294.0 ± 34.2	5.495	0.839	135
F1	15	239.40 ± 25.4	9.276	0.952	152
F2	15	198.40 ± 14.6	13.415	0.951	145
F3	15	183.40 ± 16.6	10.352	0.872	122
PTN					
X1	15	334.69 ± 67.5	5.062	0.865	154
X2	15	176.93 ± 32.3	5.221	0.950	78
X3	15	133.27 ± 31.5	4.274	0.781	49

SD: standard deviation. Weibull calculations included the Weibull modulus (m), the coefficient of determination (R-squared), and the predicted time in seconds for 99% survivability.

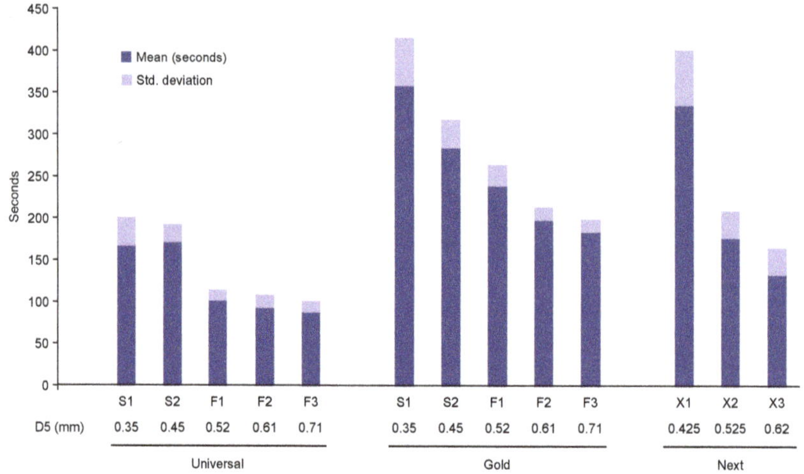

Figure 4. The mean time to fracture (s), standard deviation (SD) and D5 (mm) for PTU, PTG, and PTN.

Probabilistic modeling of fatigue failure and reliability assessment has been done for various engineering components such as turbine blades [19], turbine disc [20], and railway axles [21], which are subjected to variable loading conditions. Reliability analysis is important for the establishment of suitable safety levels for any device or system.

Weibull reliability analysis results and the probabilities of survival calculated for the PTU, PTG, and PTN instruments are presented in Table 4. The PTG series showed higher reliability than the PTU and PTN series. PTG S1 showed the longest resistance, with 181 s at 99% survival. Regarding the instruments with similar diameters at 5 mm from the tip, rotation for 152 s was predicted for PTG F1 at 99% reliability compared with 78 and 62 s for PTN X2 and PTU F1, respectively. Additionally, rotation for 145 s was predicted for PTG F2 at 99% reliability compared with 49 and 48 s for PTN X3 and PTU F2, respectively.

Figure 5 shows the fractography analysis of the PTU S1 sample. Two distinct regions were noticed: one with fatigue striations (Region a) and another with a dimpled surface (Region b) (Figure 5A). The crack initiates at the edge and propagates to the fatigue striations (Figure 5B). Micro-voids produced coalesce with each other and weakens the material (Figure 5C), after which ductile fracture occurs, which is evident from the dimpled surface in Figure 5D, until failure. The round dimples indicate normal rupture caused by tensile stresses.

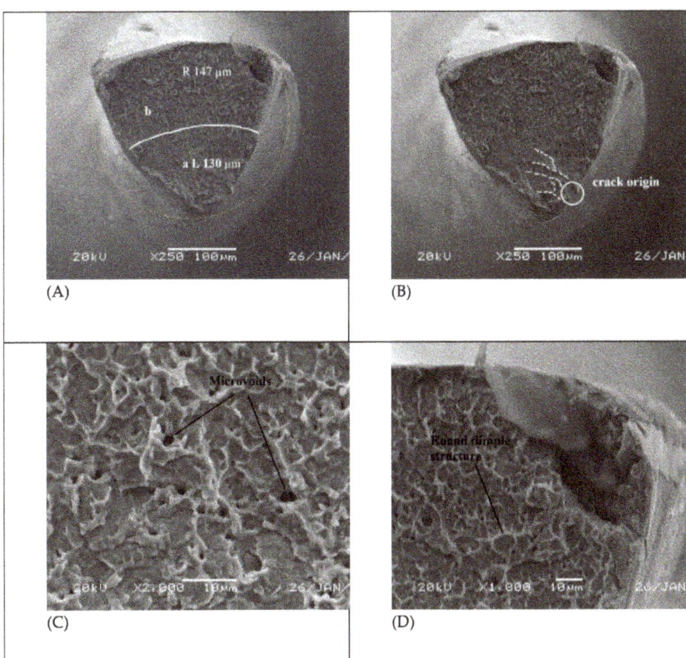

Figure 5. SEM analysis of PTU S1 sample. (**A**) Overall cross-sectional view (**B**) Crack initiation (**C**) micro-voids. (**D**) Dimpled structures.

4. Discussion

In this study, the tested instruments were selected because they shared the same recommended scheme of instrumentation. Generally, compared with the PTU and PTN series, the PTG series in this study demonstrated favorable CF behavior. However, the PTU, PTG, and PTN instruments vary in their tapering schemes, cross sections, axes of rotations, and alterations in metallurgic processing. Therefore, comparisons were performed among instruments of similar diameter (±0.01 mm) at the center of the curvature to minimize confounding factors. The PTG instruments were most resistant to CF, followed by PTN and PTU. However, the difference between PTN X3 and PTU F2 was not significant.

The higher CF resistance of the PTG and PTN instruments can be attributed to the thermomechanical treatment of these instruments [9]. Furthermore, instrument morphology is considered a significant determinant of CF behavior and can explain the greater CF resistance of the PTG instruments compared with that of the PTN instruments. Some studies have shown that instrument design is not an important determinant of CF resistance [22,23], whereas others have suggested that a different cross-sectional design is a main determinant of the CF resistance of different files [10,12,14,24,25]. Cheung et al. reported that instruments with a triangular cross-section demonstrated a higher fatigue resistance than those with a square cross section [26].

Yong et al. reported a favorable balanced relationship of flexibility, peak torque, and cyclic fatigue resistance of NiTi rotary instruments when compared to stainless steel instruments [27]. Furthermore, the thermomechanically treated NiTi instruments demonstrated greater flexibility and fatigue resistance than conventional SE NiTi instruments of similar diameter and geometry.

Several studies have compared the CF resistance of different NiTi rotary systems. Hieawy et al. tested the CF resistance of PTG and PTU instruments of sizes S1 to F3 using a 3-point bending device at a curvature of 40° with a 6 mm radius [7]. Their results showed that the PTG file had a significantly higher CF resistance than did the PTU file ($P < 0.001$). In addition, the S1 and S2 files were more resistant to fatigue failure than the F1 to F3 files in both the PTG and PTU systems ($P < 0.001$). PTG S1 showed the highest CF resistance among all files ($P < 0.001$), whereas PTU F3 showed the lowest CF resistance. These findings agree with the results from the PTU and PTG series in this study.

Perez-Higueras et al. reported that PTN X2 and X3 were more resistant than were PTU F1 and F2, respectively, when tested in stainless-steel curved canals at a curvature of 60° with a 3 mm radius [28]. Nguyen et al. compared the CF among PTU and PTN instruments with a curvature of 90° and a 5 mm radius. Their results indicated that PTU F1/F2 had a higher CF resistance than PTN X2/X3, respectively; however, this difference was not significant [29].

Weibull analysis can predict a product's resistance and be used to compare the reliability of various product designs. Nguyen et al. discussed in detail the clinical relevance and the advantages of Weibull analysis in such cases when they compared CF with PTN, PTU, and Vortex Blue rotary instruments [29]. Weibull prediction can also provide the clinician with information about the time required for a rotating instrument in a canal to fracture.

Topographic features of the fracture surfaces of all broken instruments were analyzed using SEM. The findings of this study are in agreement with previous studies where the fracture surfaces of all groups showed typical features of CF with one or more crack initiation areas, fatigue striation, and a fast fracture zone with dimples [5,7].

The KT (Kitagawa–Takahashi) diagram represents the boundary in terms of crack size and stress range for infinite fatigue life [30]. The present work can be extended to define the KT diagram for fatigue life prediction based on various approaches such as the probabilistic S-N model, EIFS (equivalent initial flaw size), and fatigue crack growth models [18,31,32]. The experimental fatigue life data can then be plotted and compared with the KT diagram.

5. Conclusions

Compared with the PTU and PTN series, the PTG series demonstrated superior CF behavior. The PTU series was the least resistant to CF. As fatigue resistance is one of the most common factors attributed to instrument fracture and considering the recent advances in technology, international standards for CF testing devices should be established to minimize the variations in reported data. Furthermore, the experimental results and SEM analysis used in this work can be used to define the KT diagram for fatigue life prediction.

Acknowledgments: The authors thank King Saud University, College of Dentistry Research Center (CDRC), particularly the physical laboratory staff, for their assistance in conducting this study.

Author Contributions: Abdullah Alqedairi and Hussam Alfawaz designed the study, supervised and performed the experiments, analyzed the data, and finalized the manuscript. Amani bin Rabba, Areej Almutairi, and Sarah Alnafaiy performed the experiments and wrote the initial draft of manuscript. Muneer Khan Mohammed designed and fabricated the artificial canals, helped in statistical analysis and wrote the laser micromachining part of the manuscript.

Conflicts of Interest: The authors have no conflicts of interest to declare.

References

1. Walia, H.; Brantley, W.A.; Gerstein, H. An initial investigation of the bending and torsional properties of nitinol root canal files. *J. Endod.* **1988**, *14*, 346–351. [CrossRef]
2. Hülsmann, M.; Peters, O.A.; Dummer, P.M.H. Mechanical preparation of root canals: Shaping goals, techniques and means. *Endod. Top.* **2005**, *10*, 30–76. [CrossRef]
3. Peters, O.A. Current challenges and concepts in the preparation of root canal systems: A review. *J. Endod.* **2004**, *30*, 559–567. [CrossRef] [PubMed]
4. Nitinol and Stainless Steel: Property Comparison. Available online: http://jmmedical.com/resources/231/Comparison-of-Properties-of-NiTi-and-Stainless-Steel.html (accessed on 28 December 2017).
5. Elnaghy, A.M.; Elsaka, S.E. Assessment of the mechanical properties of protaper next nickel-titanium rotary files. *J. Endod.* **2014**, *40*, 1830–1834. [CrossRef] [PubMed]
6. Pruett, J.P.; Clement, D.J.; Carnes, D.L., Jr. Cyclic fatigue testing of nickel-titanium endodontic instruments. *J. Endod.* **1997**, *23*, 77–85. [CrossRef]
7. Hieawy, A.; Haapasalo, M.; Zhou, H.; Wang, Z.J.; Shen, Y. Phase transformation behavior and resistance to bending and cyclic fatigue of protaper gold and protaper universal instruments. *J. Endod.* **2015**, *41*, 1134–1138. [CrossRef] [PubMed]
8. Plotino, G.; Grande, N.M.; Cordaro, M.; Testarelli, L.; Gambarini, G. A review of cyclic fatigue testing of nickel-titanium rotary instruments. *J. Endod.* **2009**, *35*, 1469–1476. [CrossRef] [PubMed]
9. Shen, Y.; Zhou, H.M.; Zheng, Y.F.; Peng, B.; Haapasalo, M. Current challenges and concepts of the thermomechanical treatment of nickel-titanium instruments. *J. Endod.* **2013**, *39*, 163–172. [CrossRef] [PubMed]
10. Haikel, Y.; Serfaty, R.; Bateman, G.; Senger, B.; Allemann, C. Dynamic and cyclic fatigue of engine-driven rotary nickel-titanium endodontic instruments. *J. Endod.* **1999**, *25*, 434–440. [CrossRef]
11. Ruddle, C.J.; Machtou, P.; West, J.D. Endodontic canal preparation: Innovations in glide path management and shaping canals. *Dent. Today* **2014**, *33*, 118–123. [PubMed]
12. Plotino, G.; Grande, N.M.; Sorci, E.; Malagnino, V.A.; Somma, F. A comparison of cyclic fatigue between used and new mtwo Ni-Ti rotary instruments. *Int. Endod. J.* **2006**, *39*, 716–723. [CrossRef] [PubMed]
13. Plotino, G.; Grande, N.M.; Sorci, E.; Malagnino, V.A.; Somma, F. Influence of a brushing working motion on the fatigue life of NiTi rotary instruments. *Int. Endod. J.* **2007**, *40*, 45–51. [CrossRef] [PubMed]
14. Gambarini, G.; Grande, N.M.; Plotino, G.; Somma, F.; Garala, M.; De Luca, M.; Testarelli, L. Fatigue resistance of engine-driven rotary nickel-titanium instruments produced by new manufacturing methods. *J. Endod.* **2008**, *34*, 1003–1005. [CrossRef] [PubMed]
15. Mohammed, M.K.; Al-Ahmari, A.; Umer, U. Multiobjective optimization of Nd: YAG direct laser writing of microchannels for microfluidic applications. *Int. J. Adv. Manuf. Technol.* **2015**, *81*, 1363–1377. [CrossRef]
16. Montgomery, D.C. *Design and Analysis of Experiments*; John Wiley & Sons: New York, NY, USA, 2017; ISBN 978-1-119-11347-8.
17. De Oliveira Correia, J.A.F.; Pedrosa, B.A.S.; Raposo, P.C.; De Jesus, A.M.P.; dos Santos Gervásio, H.M.; Lesiuk, G.S.; da Silva Rebelo, C.A.; Calçada, R.A.B.; da Silva, L.A.P.S. Fatigue Strength Evaluation of Resin-Injected Bolted Connections Using Statistical Analysis. *Engineering* **2017**. [CrossRef]
18. Correia, J.A.F.O.; Blasón, S.; De Jesus, A.M.P.; Canteli, A.F.; Moreira, P.M.G.P.; Tavares, P.J. Fatigue life prediction based on an equivalent initial flaw size approach and a new normalized fatigue crack growth model. *Eng. Fail. Anal.* **2016**, *69*, 15–28. [CrossRef]
19. Zhu, S.-P.; Huang, H.-Z.; Peng, W.; Wang, H.-K.; Mahadevan, S. Probabilistic Physics of Failure-based framework for fatigue life prediction of aircraft gas turbine discs under uncertainty. *Reliab. Eng. Syst. Saf.* **2016**, *146*, 1–12. [CrossRef]

20. Zhu, S.P.; Liu, Q.; Lei, Q.; Wang, Q. Probabilistic fatigue life prediction and reliability assessment of a high pressure turbine disc considering load variations. *Int. J. Damage Mech.* **2017**, 1056789517737132. [CrossRef]
21. Zhu, S.P.; Huang, H.Z.; Li, Y.; Liu, Y.; Yang, Y. Probabilistic modeling of damage accumulation for time-dependent fatigue reliability analysis of railway axle steels. *Proc. Inst. Mech. Eng. Part F J. Rail Rapid Transit* **2015**, *229*, 23–33. [CrossRef]
22. De Melo, M.C.C.; de Azevedo Bahia, M.G.; Buono, V.T.L. Fatigue resistance of engine-driven rotary nickel-titanium endodontic instruments. *J. Endod.* **2002**, *28*, 765–769.
23. Cheung, G.S.; Darvell, B.W. Low-cycle fatigue of NiTi rotary instruments of various cross-sectional shapes. *Int. Endod. J.* **2007**, *40*, 626–632. [CrossRef] [PubMed]
24. Ray, J.J.; Kirkpatrick, T.C.; Rutledge, R.E. Cyclic fatigue of endosequence and K3 rotary files in a dynamic model. *J. Endod.* **2007**, *33*, 1469–1472. [CrossRef] [PubMed]
25. Tripi, T.R.; Bonaccorso, A.; Condorelli, G.G. Cyclic fatigue of different nickel-titanium endodontic rotary instruments. *Oral Surg. Oral Med. Oral Pathol. Oral Radiol. Endod* **2006**, *102*, e106–e114. [CrossRef] [PubMed]
26. Cheung, G.S.; Zhang, E.W.; Zheng, Y.F. A numerical method for predicting the bending fatigue life of NiTi and stainless steel root canal instruments. *Int. Endod. J.* **2011**, *44*, 357–361. [CrossRef] [PubMed]
27. Gao, Y.; Gutmann, J.L.; Wilkinson, K.; Maxwell, R.; Ammon, D. Evaluation of the impact of raw materials on the fatigue and mechanical properties of profile vortex rotary instruments. *J. Endod.* **2012**, *38*, 398–401. [CrossRef] [PubMed]
28. Perez-Higueras, J.J.; Arias, A.; de la Macorra, J.C.; Peters, O.A. Differences in cyclic fatigue resistance between protaper next and protaper universal instruments at different levels. *J. Endod.* **2014**, *40*, 1477–1481. [CrossRef] [PubMed]
29. Nguyen, H.H.; Fong, H.; Paranjpe, A.; Flake, N.M.; Johnson, J.D.; Peters, O.A. Evaluation of the resistance to cyclic fatigue among protaper next, protaper universal, and vortex blue rotary instruments. *J. Endod.* **2014**, *40*, 1190–1193. [CrossRef] [PubMed]
30. Kitagawa, H.; Takahashi, S. Applicability of Fracture Mechanics to Very Small Cracks or the Cracks in the Early Stages. In Proceedings of the Second International Conference on Mechanical Behavior of Materials, American Society for Metals, Metals Park, OH, USA, 16–20 August 1976; p. 627.
31. Correia, J.; De Jesus, A.; Fernández-Canteli, A.; Brighenti, R.; Moreira, P.; Calçada, R. A procedure to obtain the probabilistic kitagawa-takahashi diagram. *UPB Sci. Bull. Ser. Mech. Eng.* **2016**, *78*, 3–12.
32. Maierhofer, J.; Gänser, H.-P.; Pippan, R. Modified Kitagawa–Takahashi diagram accounting for finite notch depths. *Int. J. Fatigue* **2015**, *70*, 503–509. [CrossRef]

metals
MDPI

Article

Failure Analysis of PHILOS Plate Construct Used for Pantalar Arthrodesis Paper I—Analysis of the Plate

Jason Ina [1], Madhurima Vallentyne [2], Farah Hamandi [2], Kathleen Shugart [3], Michael Boin [4], Richard Laughlin [4] and Tarun Goswami [2,4,*]

1 Boonshoft School of Medicine, Wright State University, Dayton, OH 45435, USA; ina.2@wright.edu
2 Department of Biomedical, Industrial, and Human Factors Engineering, Wright State University, Dayton 45435, OH, USA; vallentyne.2@wright.edu (M.V.); hamandi.3@wright.edu (F.H.)
3 Materials Characterization Facility, UES Inc., 4401 Dayton-Xenia Road, Beavercreek, OH 45432, USA; kathleen.cissel.ctr@us.af.mil
4 Department of Orthopedic Surgery, Sports Medicine and Rehabilitation, Wright State University, Dayton, OH 45435, USA; michael.boin@wright.edu (M.B.); richard.laughlin@wright.edu (R.L.)
* Correspondence: tarun.goswami@wright.edu; Tel.: +1-937-775-5120

Received: 21 December 2017; Accepted: 8 March 2018; Published: 13 March 2018

Abstract: The failure of a proximal humerus internal locking system (PHILOS) used in a pantalar arthrodesis was investigated in this paper. PHILOS constructs are hybrids using locking and non-locking screws. Both the plate and the screws used in the fusion were obtained for analysis. However, only the plate failure analysis is reported in this paper. The implant had failed in several pieces. Optical and scanning electron microscopic analyses were performed to characterize the failure mode(s) and fracture surface. The chemical composition and mechanical properties of the plate were determined and compared to controlling specifications to manufacture the devices. We found that equivalent tensile strength exceeded at the locations of high stress, axial, and angular displacement and matched the specification at the regions of lower stress/displacement. Such a region-wise change in mechanical properties with in vivo utilization has not been reported in the literature. Evidence of inclusions was qualitatively determined for the stainless steel 316L plate failing the specifications. Pitting corrosion, scratches, discoloration and debris were present on the plate. Fracture surface showed (1) multi-site corrosion damage within the screw holes forming a 45° maximum shear force line for crack-linking, and (2) crack propagation perpendicular to the crack forming origin that may have formed due to the presence of inclusions. Fracture features such as beach marks and striations indicating that corrosion may have initiated the crack(s), which grew by fatigue over a period of time. In conclusion, the most likely mechanism of failure for the device was due to corrosion fatigue and lack of bony in-growth on the screws that may have caused loosening of the device causing deformity and pre-mature failure.

Keywords: pantalar; arthrodesis; failure modes; fracture surface; corrosion; scratching

1. Introduction

Arthrodesis has served as a standard treatment for painful arthritis of the ankle and the subtalar joints. The ankle joint is composed of the tibia and fibula and the subtalar joint includes the talus and calcaneus [1]. Fusion of the ankle, subtalar, and talonavicular joints together is known as pantalar arthrodesis and has been performed by orthopaedic surgeons since 1906 [2]. There also are non-surgical treatment methods such as non-steroidal anti-inflammatory medications and prefabricated or custom orthotics, but these methods are not effective in all patients.

Pantalar arthrodesis can be achieved with internal or external fixation. Internal fixation is preferred in most cases as it has many advantages over external fixations. The incidence of non-unions and mal-unions are reduced with internal fixation [3]. Implants that can be used for internal fixation are plates, screws, intramedullary nails, and wires. Locking compression plates (LCP) are advantageous when compared to others since they provide an angular-stable interface. LCPs are most commonly used for acute traumatic fractures, osteotomy fixation, non-union repair, and arthrodesis [4]. In the compression plating technique with conventional screws, primary stability is achieved by the compression between the plate and the bone. However, since there is no direct contact between the plate and the bone, the primary stability achieved by compression may be lost due to cyclic loading. The two major mechanisms for fixation failure in these constructs are axial tilting in the absence of the normal support given by bone due to fracture comminution and screw loosening in osteoporotic bone. In these cases, locking screws are preferred to conventional screw to limit these mechanisms of failure. Because locking screws lock into the locking plate, the amount of axial tilt and screw loosening is minimized. Proximal humerus internal locking system (PHILOS) (Figure 1) plating used in the pantalar joint utilizes the benefit that blood supply to the bone is preserved and soft-tissue injury minimized due to lack of compression. In addition, successful fixation result in anatomic reduction, stable fixation, preservation of blood supply, and early mobilization [5–9] which occur with the use of locking compression technology.

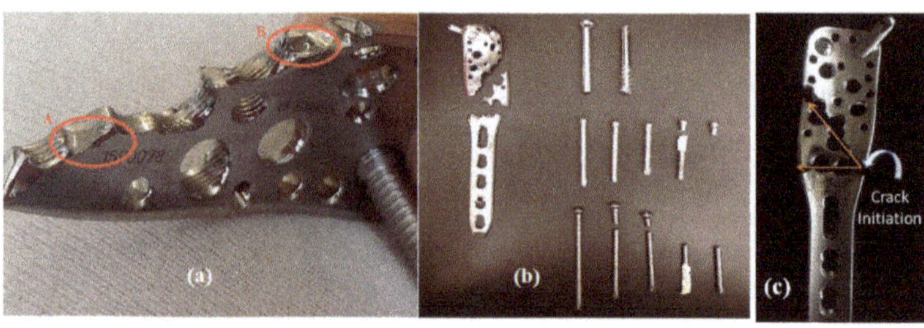

Figure 1. Submitted proximal humerus internal locking system (PHILOS) device for analysis; (**a**) plate through the thickness shows regions of interests marked by A and B and 1–3 investigated in this paper (**b**) the failed samples and (**c**) crack initiation and propagation of the plate.

A summary of device failures was compiled from literature [10–23] (see Appendix A). It identifies the devices, material of construction, failure modes and other fracture features. It is noteworthy that this may be the only case involving the failure analysis of PHILOS (Figure 1). A number of studies involving SS316L material revealed inclusion sites as crack origins [10,15–18], corrosion, wear and fatigue [10–13,15,16,18,19,21,23] with no visible mechanical failure yet having failed clinically [12,14]. The majority of the devices were constructed with SS 316 and 316L; however, a few devices constructed with Ti-6Al-4V alloy [19–22] and pure titanium [23] were also found. From this summary it is prudent to investigate the chemical composition and inclusions that form pitting and corrosion fatigue related failures of the devices.

1.1. Case Presentation (Clinical Summary)

1.1.1. Initial Presentation

A 68-year-old female was seen in the office for a right foot deformity. She experienced changes in ambulation over the past month with no history of trauma. Other than her progressive ankle deformity, she was relatively healthy.

On physical exam there was a fixed, non-reducible deformity of the right ankle. Palpable posterior tibial and dorsalis pedis pulses were present. Sensation over the dorsal and plantar foot was intact to light touch. Initial ankle X-rays were significant for a right ankle deformity consistent with chronic lateral subtalar dislocation with varus tibiotalar joint alignment (Appendix B, Figures A1 and A2).

There was no evidence of bone loss or acute fracture. It was determined that the patient required a right ankle pantalar arthrodesis.

1.1.2. Operation

A 3.5 mm LCP PHILOS (Synthes, Solothurn, Switzerland) was inverted so that the locking screws could be inserted into both the talus and calcaneus and compression screws into the tibia. The plate was fixed to the lateral portion of the calcaneus and tibia using four 3.5 mm cortical screws proximally in the tibia and seven 3.5 mm locking screws distally. A 6.5 cannulated screw (Zimmer, Warsaw, IN, USA) was placed from the tibia across the talus and into the calcaneus to provide compression across the tibiotalar and subtalar joints. Three additional 3.5 mm screws with washers were placed from the navicular into the talus in order to fuse the talonavicular joint.

An additional 3.5 mm screw and washer were inserted from the anterior process of the calcaneus to the cuboid to achieve fusion at this joint. The patient was to remain non-weight bearing on her surgical extremity for three months.

1.1.3. Post-Operative Follow-Up

Clinically, in the first year of follow up appointments, the patient was doing very well. Weight bearing as tolerated began after three months. She returned to the surgeon's office once yearly for follow up. During every follow up, new imaging (by X-ray) of the ankle was obtained. Imaging over this time showed that her ankle did not fuse and formed a non-union. Her hardware failed as seen with radiographic evidence of breakage. The hardware breakage progressed from a single screw to involving multiple screws and the plate over the post-operative course. At six years post-operative she complained of a new onset pain in her operative ankle. With the new onset symptoms and evidence of failed fusion and hardware failure, a removal of hardware and revision ankle fusion was planned.

1.1.4. Post-Operative Imaging

X-rays from the 18-month post-op visit showed evidence of breakage of a distal 3.5 mm talonavicular screw (Figure A2). The fracture line was located in the thread near the screw head. At two years post-op, the screw breakage was more displaced and there was an additional fracture in the most proximal 3.5 mm cortical screw in the tibia (Figure A3). The fracture line of this screw was again in the thread near the screw head. At three years post-op, a new fracture of the most proximal 3.5 mm locking screw in the talus was present (Figure A4). There was also evidence of a newly forming valgus deformity. At five years post-op, an additional 3.5 mm cortical screw in the tibia, a second 3.5 mm talonavicular screw, the 6.5 mm cannulated screw, and the 3.5 mm LCP plate all exhibited fracture on imaging (Figure A5). At six years post-op, an additional 3.5 mm cortical screw in the tibia broke and the fractured plate became more widely displaced (Figure A6). The valgus deformity seen earlier had now progressed to nearly 30 degrees.

2. Materials and Methods

The PHILOS plate was made of Stainless Steel 316L, originally manufactured by Synthes, now Depuy Synthes Joint Reconstruction (Warsaw, IN, USA). The plate is anatomically shaped to the complex contours of the proximal humerus. The construct contains both locking and combination of locking and compression holes using different types of screws [24,25]. There are 10 locking holes in the distal end of the PHILOS permitting multiple points of fixation for support. The PHILOS has 5 combination holes, one elongated hole to aid in-plate positioning with a 3.5 mm locking screws hole in the threaded portion in the proximal shaft. There are also 3.5 and 4.0 mm cortical screw holes and 4.0 mm cancellous bone screws in the compression portion. All 18 pieces of the implant submitted for investigation are shown in Figure 1. Among the submitted pieces, only 4 screws were intact. The LCP, which was fractured into three pieces, was measured to be 114 mm in length and had a thickness of 3.5 mm.

In order to assess the mechanical and manufacturing integrity of the failed device, we preformed Rockwell hardness B-scale test and X-ray energy dispersive spectroscopy (EDS) using a Quanta 600 scanning electron microscope (SEM, Thermo Fisher Scientific, Hillsboro, OR, USA) with a 15 kV voltage and a spot size of 5. We compared the results from our testing to ASTM (American Society for Testing and Materials) standards reported for stainless steel 316L [26]. The Rockwell hardness test was performed in three different locations of the fractured plate with each test location increasing in distance away from the fracture surface. The EDS was performed using a 10 mm square detector and genesis software package. EDS was performed on both the fracture plate. EDS samples were taken both at the plate surface and plate interior to ensure homogeneity throughout both pieces.

One of the plate fragments was cleaned in ethanol followed by water sonication for 10 min to study the fracture surface. Fractography was performed using the SEM with a voltage of 15 kV and a spot size of 3. Electron backscatter diffraction (EBSD) was performed on an FEIXL-30 using an EDAX EBSD detector to determine the grain size and orientation.

3. Results and Discussion

3.1. Chemical Composition Characterization

The results of the X-ray EDS for the fracture plate is shown in Table 1 and compared with reported ASTM standard for SS 316L stainless steel (F138-03). An additional requirement set forth by the ASTM standard is shown in Equation (1), which yielded 24.77 for the plate [26]. One factor to keep in mind when comparing EDS results to the ASTM standard is that EDS is known to be a semi-quantitative technique. In addition, adventitious carbon may have accumulated on the surface preventing C from being quantified. There is also an overlap between the X-ray peaks of Mo and S, precluding deconvolution of their concentrations. The EDS system used in this case was not sensitive enough to detect the trace amounts of P, S, and N expected according to the standard. The EDS spectrum is shown in Figure 2.

$$\%\,\text{Cr} + 3.3 \times \%\text{Mo} \geq 26.0 \quad (1)$$

Table 1. X-Ray energy dispersive spectroscopy (EDS) measured composition for plate (wt %).

Composition	ASTM Standard (Max)	Plate (Average)	Tolerance Limits
C	0.03	-	0.005
Mn	2.0	1.81	0.04
Si	0.75	0.59	0.05
P	0.025	-	0.005
S	0.01	-	0.005
Cr	17–19	17.61	0.05
Mo	2.25–3.0	2.17	0.15
Ni	13–15	13.84	0.2
N	0.10	-	0.01
Fe	Balance	63.99	-

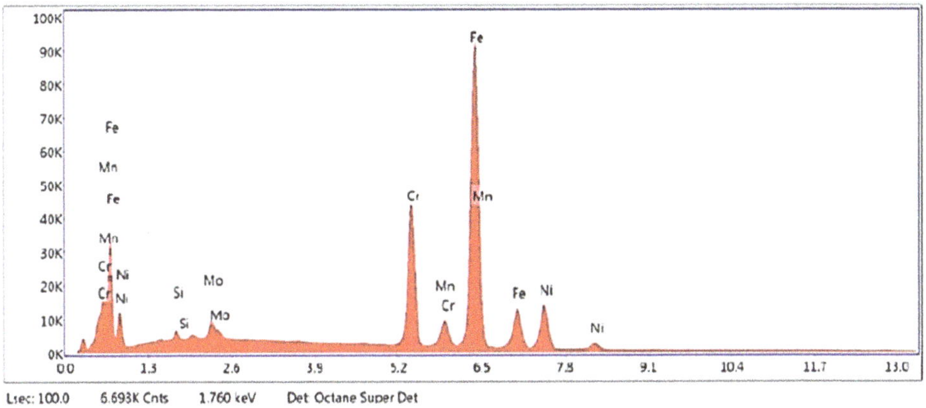

Figure 2. Energy dispersive X-ray analysis shows qualitatively the peaks of different element weight percent present in the PHILOS plate.

The plate conformed to ASTM standards according to the tolerated ranges of element composition but failed equation 1 in this instance shown in Table 1. However, in another instance, the results of the composition analysis gave a result of 28.7 for Equation (1), thus passing the requirements.

3.2. Microstructural Characterization

EBSD and backscatter imaging methods were used to characterize the structure of the material. The microstructure is shown in Figure 3. The range in the average grain size of the plate was four to six micrometers shown in Figure 4.

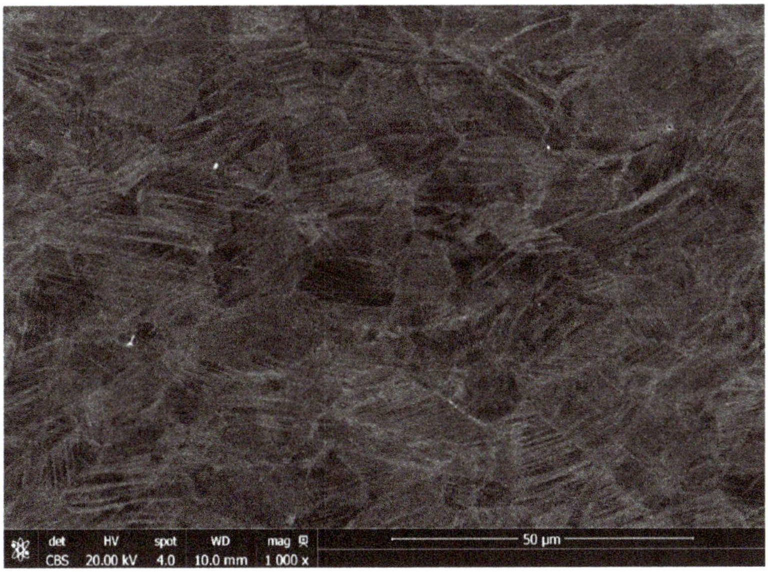

Figure 3. Microstructure of PHILOS plate at 1000×.

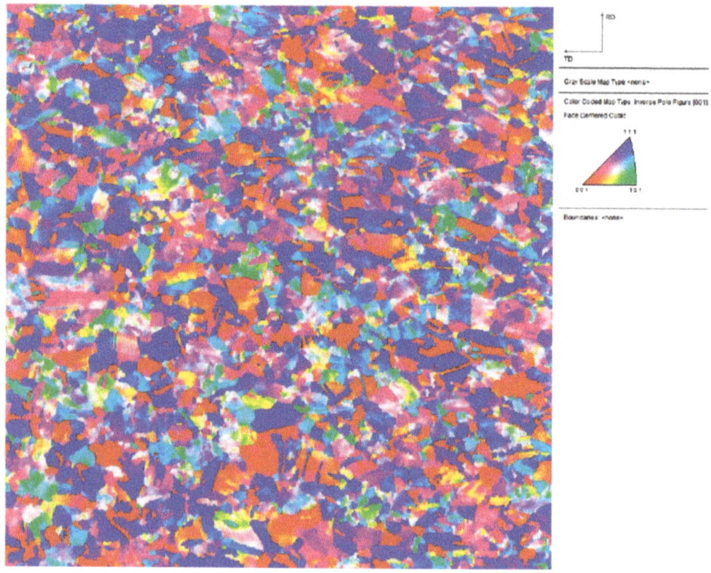

Figure 4. Electron backscatter diffraction (EBSD) grain map of the PHILOS plate.

An attempt was made to identify and characterize the inclusions. Two regions of interest identified in Figure 1 as area 2 and 3, were investigated for inclusions and presented in Figure 5a,b. The composition of one inclusion was mapped in Figure 6 using EDS. The maps in Figure 6 show that inclusions are likely Mo rich (though this could be S), these inclusions may be fine, localized chi and sigma intermetallic phases which may have started the crack. There is a need to characterize inclusions in surgical grade SS 316L as a dedicated basic research focus since this paper deals with failure analysis and the importance that the inclusions have on failure initiation.

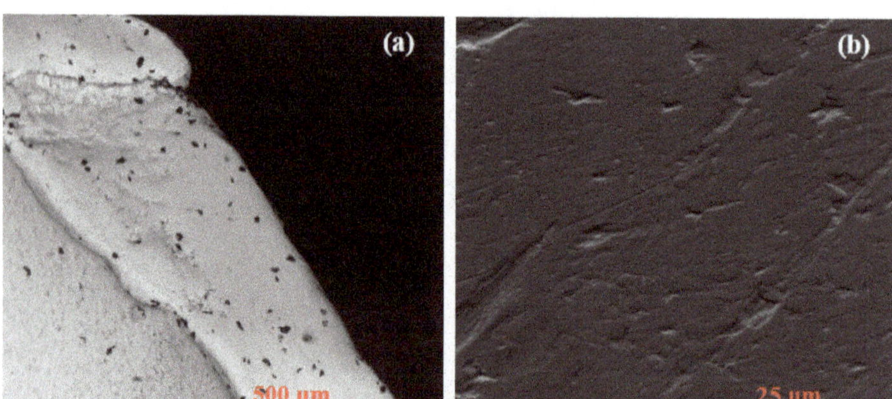

Figure 5. Backscattered electron images of likely inclusions (**a**) SEM EBSD image of area 2 (**b**) Topographical SEM image of area 3.

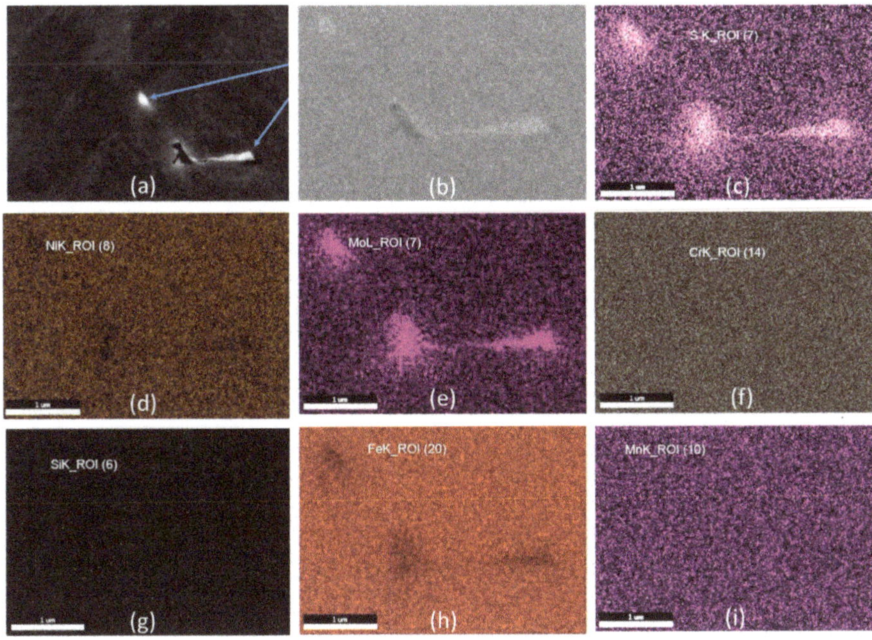

Figure 6. Characterization of inclusions using EDS. (**a**) The inclusions, (**b**) Gray scale of inclusions. Overlapping spectra of S and Mo visible in (**c**) Sulfur and/or (**d**) Nickel, (**e**) Molybdenum in the microstructure of the PHILOS plate. While the rest (**f**) Chromium, (**g**) Silicon, (**h**) Iron and (**i**) Manganese are not visible.

3.3. Material Property Analysis

The results of the Rockwell hardness B-scale test on the fractured plate are shown in Table 2. The ASTM standard upper limit for SS 316L is 95 HRB, and a 95% confidence interval was calculated for each trial. The hardness was then converted to Vickers hardness and tensile strength, and a 95% confidence interval was also calculated and used to compare against the ASTM standard for SS316L [26]. Hardness of the plate varied across the plate. It is evident that as the fatigue related deformation accrued, the material hardened across or near the fracture plane. It is likely that the material may have been supplied at higher strength conditions via cold working and may also have had reduced elongation.

Table 2. Rockwell hardness B-scale test, Vickers hardness, and tensile strength results.

The Test	Average	ASTM Standard	Proximal	Middle	Distal
Rockwell hardness	Average	95	101.8	107.7	107.8
	95% confidence interval	-	101.1–102.5	106.1–109.3	106.6–109.5
Vickers hardness	Average	213	266.5	330	334.5
	95% confidence interval	-	260–273	310–350	316–353
Tensile strength (psi)	Average (MPa)	868	848	1057	1057
	95% confidence interval	-	837–859	1004–1110	1004–1110

3.4. Optical Microscopic Analysis

Visual observation indicated a large number of scratches on the surface of the plate. Two broken screw heads (with StarDrive recess) remained in the plate, while one partially intact locking screw was not removed from the distal end. Figure 1 shows that the crack initiated from under one of the locking screw holes near the middle of the plate. This fracture initiation has progressed in two different directions, one diagonally 45° in the directions of maximum shear and one perpendicularly (Figure 1), which caused the plate to fail into three pieces. The observed features are summarized in Table 3. Since the plate and screw construct was removed after it had failed, we do not know whether the material dissolution or contact between two mating surfaces during removal removed the pitting and or other localized damage sites prior to crack formation.

Table 3. Various damage features on different parts of implant.

Location	Damage Feature(s)
Distal area (Figure 1B)	Scratches and pits
Close to fracture plane (Figure 1A)	Scratches, pits, corrosion, cracking
Area (A), (Figure 1)	Debris, discoloration

3.5. Fractography

The markings shown in Figure 7a are changes in surface elevation eluding to pitting in area of interest 3. Figure 5 shows EBSD image and distribution of different phases present in the material that may be described as non-metallic/inclusions. Non-metallic inclusions are known to be areas of stress concentration and increase the likelihood of pitting corrosion and secondary crack initiation. Sudhakar et al. and Kanchanomi et al. found that non-metallic inclusions initiated the eventual failure of devices in their studies [13,18]. This evidence of pitting and foreign body inclusions, seen in Figures 5 and 6, respectively, supports secondary crack initiation due to corrosion-fatigue at these sites. Figure 8a–c show fatigue striations in the areas of interest 1, 2, and 3, respectively. Striations are evident within all three images. Figure 8a also shows some evidence of cleavage and facets. The striation spacing in each image was measured and displayed in Table 4. The arrows indicate the direction of crack propagation based on increase in striation spacing.

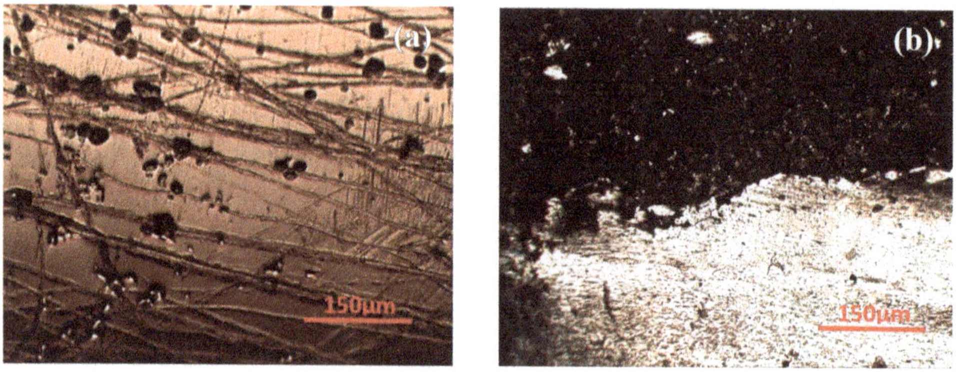

Figure 7. Light microscopy images showing (**a**) scratches and uniform pitting, and (**b**) debris and discoloration.

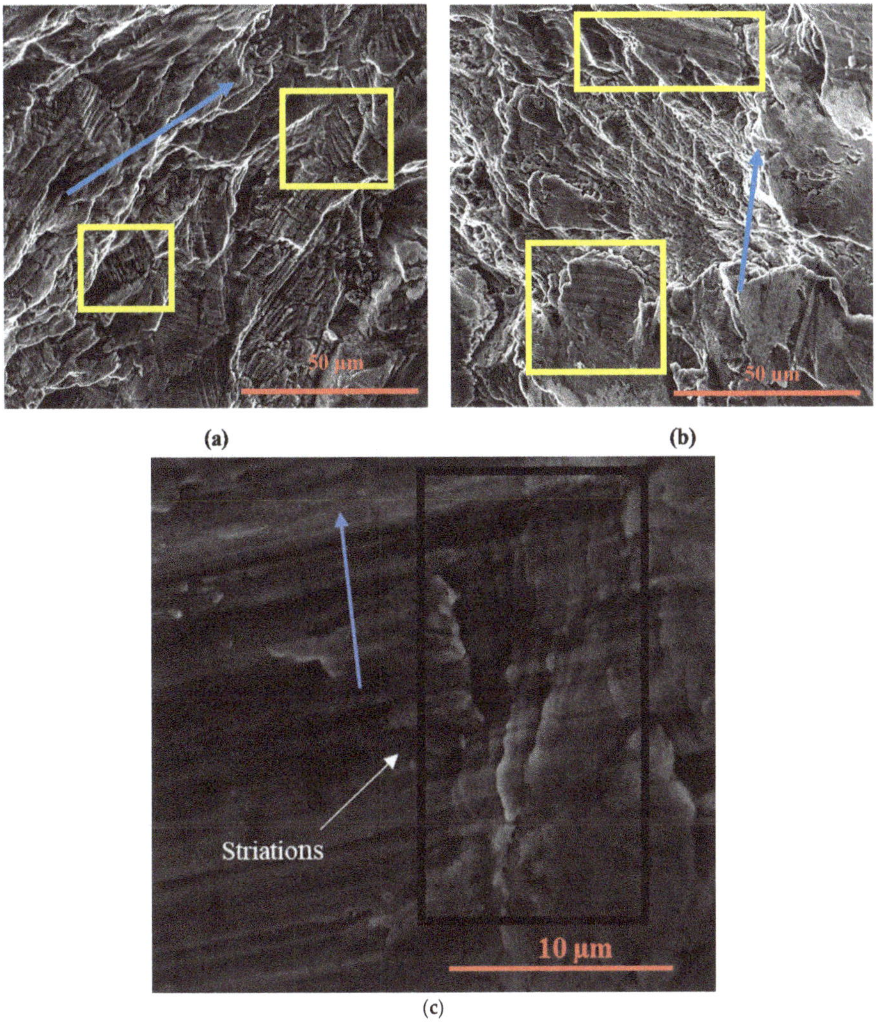

Figure 8. SEM images of (**a**) area 1, (**b**) area 2, and (**c**) area 3 showing striations perpendicular to crack propagation. The yellow boxes show striations and blue arrows denote the direction of crack propagation.

Table 4. Striation spacing range.

Reference Image	Striation Spacing (μm)
Figure 8c (area 3)	0.44–0.99
Figure 8b (area 2)	0.80–1.42
Figure 8a (area 1)	1.91–4.95

From these methods it is quite clear that striation spacing increased as the crack advanced through the three regions, 3 to 1, respectively. The origin for crack appears to be from area 3 (distal end of plate, latterly); however, as it advanced, the linking occurred through the screw holes in a 45° angle which may have coincided with dorsiflexion angle or slight movement by gait. Upon failure, the plate was not able to load bear along this angle, the load transferred perpendicularly causing cracking and linking through the screw holes in that direction as well.

The SEM imaging performed on the failed implant provides a strong argument for initial corrosion-fatigue fracture followed by brittle fracture for the mechanism of failure. Table A1 shows that Thapa et al., Karmacharya et al., Majid et al. and Azevedo et al. identified corrosion as an important factor in the failure of devices [11–13,19]. Fatigue fracture has been conservatively estimated to account for 50% of all brittle fractures in manufactured products, making fatigue fracture the most prevalent initiator of brittle fracture [27]. Striations due to secondary crack propagation were evident at multiple locations on the locking compression plate (Figure 8a–c). Thapa et al. came to the conclusion that a 10 hole locking compression plate led to failure after 2035 cycles with a striation spacing of 0.3–1.72 μm [14]. Goswami et al. showed striation spacing within a range of 10–15 μm for the fracture of an IM nail that led to an estimate of less than 100 cycles before fracture [22]. Based on the findings of previous research, it may be estimated that the plate may have accumulated a quarter of a million cycles by a stable crack propagation mechanism; however, as it jumped through the screw hole(s), the striation spacing increased. At location 1, the striation spacing was wide enough that failure may have occurred in the tertiary phase of crack propagation. However, the kinetics of how an inclusion transitioned to pitting and pit to crack is not known. These findings support evidence of secondary crack initiation, within the areas of non-metallic inclusions, as shown in Figure 9. The fatigue failure mechanism for SS 316L was in terms of excessive twinning as shown in Figure 10. However, twinning may also be a standard microstructure of SS316L and the available samples were made from the failed device; we cannot conclude whether or not the twinning were original microstructure or as a result of in vivo use.

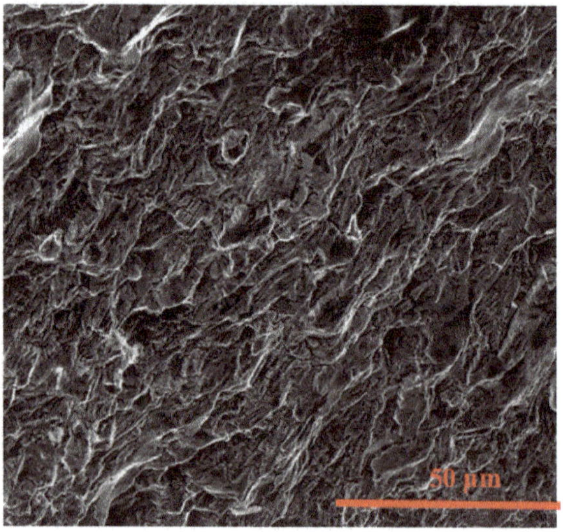

Figure 9. SEM image taken at area of interest 1. This image contains examples of quasi-cleavages and facets.

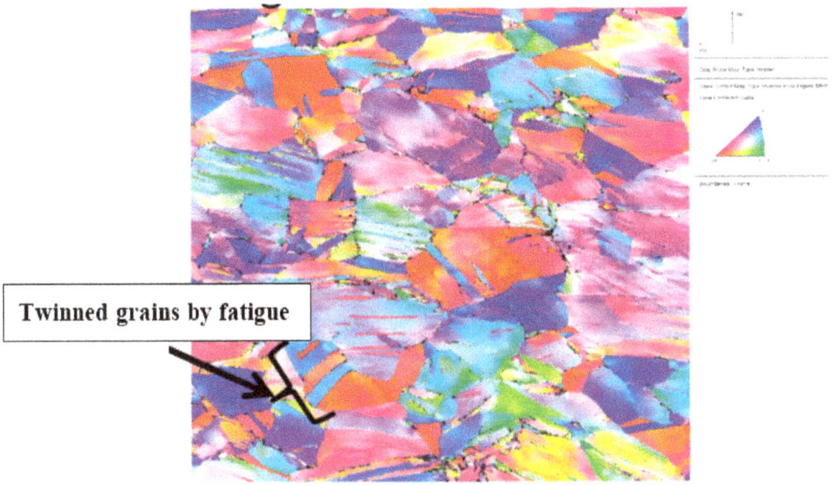

Figure 10. Inverse pole figure showing crystallographic characterization of the plate microstructure showing twinning that may have occurred due to in vivo use by fatigue or a material feature.

The quasi-cleavage and facets seen in Figure 9 are highly characteristic of brittle fracture—an end consequence of corrosion-fatigue crack propagation. Imaging of the distal end of the failed piece on the optical microscope (Figure 7b) showed evidence of progressive fatigue fracture that eventually led to brittle fracture, depicted by the abrupt change in surface pattern. It is this brittle fracture that ultimately led to the failure of the PHILOS.

Azevedo et al. investigated multiple devices (femoral compression plate, femoral nail plate, and oral maxillo-facial plate) and discussed the importance of material conforming to ISO (International Organization for Standardization) and/or ASTM standards for these devices. Most devices investigated [19] failed by a corrosion-assisted fracture that can be traced back to improper chemical composition. Other devices investigated revealed fabrication or assembly defects that contributed to their premature failure [23]. Sivakumar et al. and Marcomini et al also showed that lack of chemical conformity in the alloy had led to the implant failure [19,20]. The results from the chemical and mechanical property analysis show a potential lack of conformity to ASTM standard. However, it is also possible that the material procurement protocols may have followed other specifications as the device was then marketed by Synthes (Solothurn, Switzerland). The microstructure of the SS316L, Figure 4, shows, given the orientation of various grains via colors in the Figure, that the grains are randomly distributed indicating a weak texture. The pitting resistance of the material has not been investigated; however, is triggered by the presence of MnS inclusions, not documented for this material.

4. Conclusions

Images taken at six weeks post-op showed no evidence of device failure. The implant failed mechanically as well as clinically. Failure in the device construct caused lack of union. The first follow-up X-ray images showing evidence of device failure via screw fracture were taken at 18 months post-op. The average time for tibiotalocalcaneal arthrodesis fusion is about 19 weeks. Since imaging evidence supports device failure initiating between six weeks and 18 months, it cannot be said with certainty that the device maintained its integrity for the amount of time necessary for fusion. The revision surgery implemented a pediatric distal femoral osteotomy plate. There is no follow-up data on the revision. The following are a few reasons that may be considered for the overall failure of the PHILOS plate:

i As we know that the plate was designed for the proximal humerus, it did not match all the requirements of pantalar arthrodesis. A slight mismatch would give rise to forces sufficient to cause overloading failure.

ii The curvature in the plate is one of the important factors to be considered, due to which unequal load distribution could have taken place.

iii Improper reduction may be due to the device. This improper reduction may have generated excessive loading on the LCP leading to a premature mechanical failure.

iv The qualitative data of composition, non-metallic inclusions, hardness, and tensile strength of the plate show that the material conformity per ASTM standard may be an additional issue controlling the integrity of the device.

v SEM imaging supports a mechanism of corrosion-fatigue fracture from crack initiation sites in the distal areas of the plate due to the presence of inclusion bodies and pits. These crack initiation sites can then lead to crack propagation towards the proximal area (shown by the striations) and ultimately brittle fracture (supported by the cleavage sites and facets).

Acknowledgments: Gerard Simon, Air Force Research Laboratory, Dayton, Ohio provided access to the characterization facilities. Dinesh Gundapaneni assisted with sample preparation, microstructure and hardness testing of the material.

Author Contributions: Madhurima Vallentyne did the initial work as a part of her independent research, Jason Ina continued the further characterization work. Michael Boin reviewed the patient charts and provided clinical insights. Kathleen Shugart performed the SEM analysis and texture related research. Farah Hamandi performed mechanical tests, theoretical aspects of the paper and in the revisions. Richard Laughlin performed the revision surgery and removed the device components. Tarun Goswami, led this research study, recruited students, medical students and residents, verified the analytical methods and edited or written the body of the paper.

Conflicts of Interest: The authors declare no conflict of interest.

Appendix A

Table A1. Summary of orthopaedic device failure modes.

Study	Device Details	Alloy	Failure Mode	Details of Failure
Kanchanomai et al. [10]	14 hole, Locking compression plate (for femur)	316L Stainless steel	Excessive walking before adequate healing of fracture	- Striation spacing: 1 µm at the middle of the fracture - Thickness of LCP: 5 mm - Estimated no. cycles from crack initiation to bottom: 5000 cycles - Estimated number of cycles for complete fracture: 42,000 (8 days of walking) - Sub-surface inclusion found at initiation of crack site
Thapa et al. [11]	10 hole, Locking compression plate	Stainless steel	Corrosion-fatigue	- Striation spacing: at crack initiation—0.3 µm, at fracture surface—1.72 µm (crack length—1 mm). - Estimated no. of cycles after crack initiation: 2035 cycles

Table A1. *Cont.*

Study	Device Details	Alloy	Failure Mode	Details of Failure
				- Corrosion (rusting) visible on surface
				- Carbon intermetallic-inclusion found
Karmacharya et al. [12]	8 hole, Reconstructive locking plate	316L Stainless steel	Corrosion	- Mechanically intact & no significant damage visible
Majid et al. [13]	Lumbar plates (113 plates were studied)	316L Stainless steel	Corrosion	- 72.5% (majority) of plates showed corrosion.
				- Fe, Cr and Ni (preferential) metal ions were released into body due to corrosion
				- Corrosion occurred at screw-plate interface (fretting/crevice)
Varadharajan et al. [14]	Hemi-toe implant	Cobalt chromium alloy	Failed clinically	- Mechanically intact & no significant damage visible
				- Anterior/middle surface exhibited more scratching/debris deposition—can be indicative of damage to HMWPE bearing
				- Coat spalling may indicate device loosening
Sudhakar et al. [15]	Nail for Shinbones	316L Stainless steel	Ductile fracture and nonmetallic inclusions	- Ductile fracture facilitated by micro-fracture due to non-metallic inclusion (125 um) is predominant mode of failure
Sivakumar et al. [16]	6 hole, Tubular compression bone plate (for femur)	316L Stainless steel	Improper fixation	- Pitting potential of implant is due to higher Chromium—2.3% and Nickel—4% contents.
				- Fracture was at 5th countersunk hole of plate
				- "Beach marks" at outer edge of crack indicate torsional force
				- Grain size and inclusion content did not meet ASTM requirement
				- Bone plate was placed anteriorly rather than laterally which does not allow for proper compression
				- Screw head and screw hole mismatch

Table A1. *Cont.*

Study	Device Details	Alloy	Failure Mode	Details of Failure
Marcomini et al. [17]	Femoral Locking compression plate	Stainless steel (not 316L due to higher content of Ni and P)	Non-conformity of the material	- Phosphorus content 0.26% over superior limit - SEM showed evidence of brittle fracture due to segregation of P in grain boundaries - Steel was cold worked which contributed to brittle fracture
Cahoon et al. [18]	McLaughlin plate for hip (7 other implants discussed)	Cast Vitallium	Fracture in the area where maximum stresses were expected	- Period of time the implant served: 1 year - Fatigue strength decreased due to increase in porosity
Cahoon et al. [18]	V-Moore plate and screws	316L Stainless steel	Crevice corrosion	- Period of time the implant served: 1 year - Tissue reaction due to severe crevice corrosion
Cahoon et al. [18]	Nail plate	316L Stainless steel	Fracture due to bending (while trying to fit the patient)	- Punching of screw holes too close and too near to the edge of the nail plate - To dissolve the precipitation of chromium carbide, it was heat treated at: 1040 °C for 1 h.
Azevedo et al. [19]	13 hole reconstruction plate for osteosynthesis	Plate: CP Titanium Screws: Titanium-6Al-4V alloy	Corrosion-fatigue	- Intergranular cracking formation: 15 μm deep on notch surface - Reduction of area featured: 55% approximately. - Equiaxed α grains and Intergranular β platelets revealed - Corrosion observed on intergranular β-phase which was in contact with body fluid
De Medeiros et al. [20]	Hemimandibles with assistive 4 hole plates	5052-F Aluminum	Fracture due to ductile overload	- Number of plates fractured in the upper part of the hole to the right of the fracture line: 5 - Surface: Dim and grayish showing intense plastic deformation - Plates failed due to ductile fracture mechanism alluding to proper manufacturing

Table A1. *Cont.*

Study	Device Details	Alloy	Failure Mode	Details of Failure
Roffey et al. [21]	Femoral stem	M30NW High nitrogen Stainless steel	Failure at a point of maximum stress due to bending and torsional loading	- No. of fracture initiation points: 2
				- Large Beach marks indicate large number of cycles to failure
				- Visible ratchet marks indicative of fatigue failure
				- Scratching and scoring patterns support post-manufacturing damage
				- Fatigue fracture pattern confirms mechanical damage prior to fracture
				- Initial fracture occurred in region of lowest cross-sectional area
				- Mechanical damage to surface during implantation led to initiating fatigue crack
Goswami et al. [22]	IM nail	Titanium	Axial, bending, and torsion forces	- Interference between screw, core-lock caps, and IM nail
				- Lack of purchase to bone
				- Pre-crack due to interference
				- Striation spacing = 10–15 μm
Azevedo et al. [23]	Femoral compression plate	Stainless steel	Fretting-fatigue	- Underside of screw-head and plate hole countersink showed marks of wear and corrosion that led to removal of passivation layer
				- Chemical composition did not meet ISO 5832-I standards which decreases corrosion resistance
				- Presence of fretting corrosion may suggest screw loosening
				- Fatigue striations (spacing = 0.6 μm) associated with secondary cracking
Azevedo et al. [26]	Femoral nail-plate	Stainless steel	Fatigue fracture	- Multiple nucleation sites present at high-stress concentration points
				- Chemical composition failed to meet ISO 5832-I standards

Table A1. *Cont.*

Study	Device Details	Alloy	Failure Mode	Details of Failure
Azevedo et al. [23]	Titanium oral maxilla-facial plate	CP Titanium	Corrosion-fatigue	- Evidence of brittle fracture shown by no gross plastic deformation near fracture surface
				- Presence of fissure striations, complex furrow structures, secondary intergranular cracking
				- Intergranular cracking near surface of origin indicative of corrosion—fatigue mechanism
				- ISO 5832-2 standards were met for chemical composition of CP titanium

Appendix B

Radiological Features Showing Placement of Devices and Gradual Failures

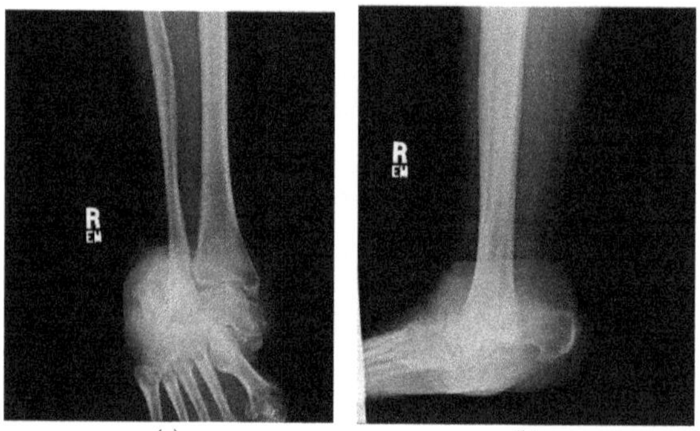

(a) (b)

Figure A1. Initial X-Ray images showing lateral subtalar dislocation with varus tibiotalar joint alignment.

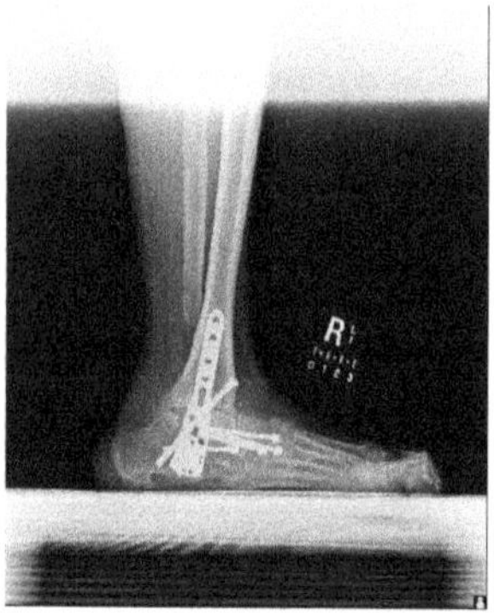

Figure A2. 18 months post-op imaging with evidence of breakage of a distal 3.5 mm talonavicular screw.

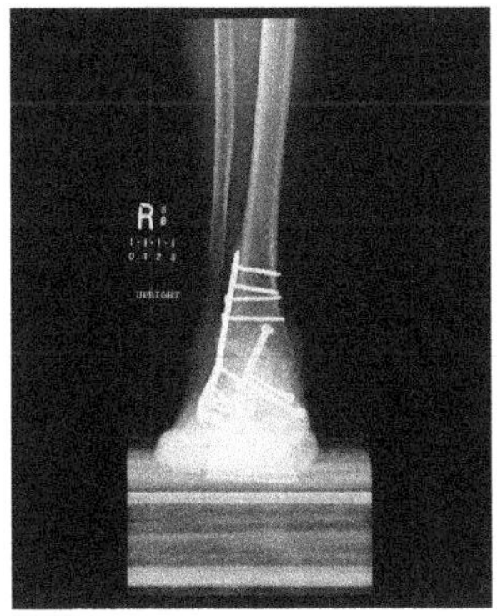

Figure A3. 2 years post-op imaging with screw fracture more displaced and an additional fracture in the proximal 3.5 mm cortical screw in the tibia.

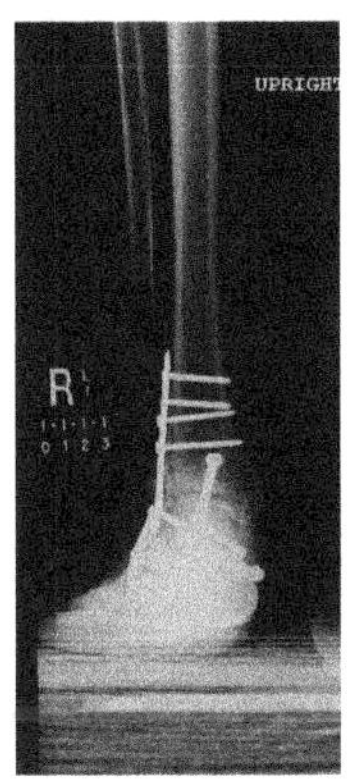

Figure A4. 3 years post-op showing new fracture of the most proximal 3.5 mm locking screw in the talus.

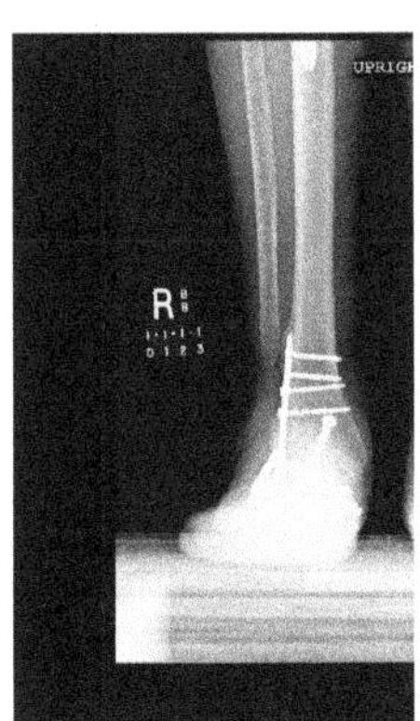

Figure A5. 5 years post-op with additional fractures in a 3.5 mm cortical screw in the tibia, a second 3.5 mm talonavicular screw, the 6.5 mm cannulated screw, and the 3.5 mm LCP plate.

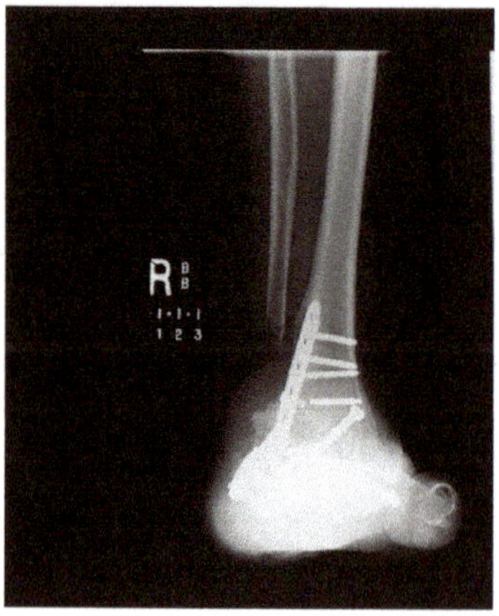

Figure A6. 6 years post-op with an additional fracture in a 3.5 cortical screw in the tibia and the fractured plate became more widely displaced. The valgus deformity seen earlier had now progressed to nearly 30 degrees.

References

1. Quill, G.E., Jr. Tibiotalocalcaneal arthrodesis. *Tech. Orthop.* **1996**, *11*, 269–273. [CrossRef]
2. Coughlin, M.J.; Nery, C.; Baumfeld, D.; Jastifer, J. Tibiotalocalcaneal Arthrodesis with lateral compression Plate. *Rev. Bras. Ortop.* **2012**, *47*, 467–473. [CrossRef] [PubMed]
3. Ortho Info. Internal Fixation for Fractures. American Academy of Orthopaedic Surgeons. Available online: https://orthoinfo.aaos.org/en/treatment/internal-fixation-for-fractures/ (accessed on 13 August 2015).
4. Sommer, C.; Gautier, E.; Müller, M.; Helfet, D.L.; Wagner, M. First clinical results of the Locking Compression Plate (LCP). *Injury* **2003**, *34*, 43–54. [CrossRef]
5. Kubiak, E.N.; Fulkerson, E.; Strauss, E.; Egol, K.A. The evolution of locked plates. *J. Bone Jt. Surg.* **2006**, *88* (Suppl. 4), 189–200.
6. Ahmad, M.; Nanda, R.; Bajwa, A.S.; Candal-Couto, J.; Green, S.; Hui, A.C. Biomechanical testing of the locking compression plate: When does the distance between bone and implant significantly reduce construct stability? *Injury* **2007**, *38*, 358–364. [CrossRef] [PubMed]
7. Desai, K. A Biomechanical Comparison of Locking Compression Plate Constructs with Plugs/Screws in Osteoporotic Bone Model. Master's Thesis, Wright State University, Fairborn, OH, USA, 2010.
8. Ahmad, J.; Pour, A.E.; Raikin, S.M. The modified use of a proximal humeral locking plate for tibiotalocalcaneal arthrodesis. *Foot Ankle Int.* **2007**, *28*, 977–983. [CrossRef] [PubMed]
9. Twaij, H.; Damany, D. PHILOS humerus plate for a distal tibial fracture. *J. Surg. Case Rep.* **2013**, *2013*, rjs036. [CrossRef] [PubMed]
10. Kanchanomai, C.; Phiphobmongkol, V.; Muanjan, P. Fatigue failure of an orthopedic implant—A locking compression plate. *Eng. Fail. Anal.* **2008**, *15*, 521–530. [CrossRef]
11. Thapa, N.; Prayson, M.; Goswami, T. A failure study of a locking compression plate implant. *Case Stud. Eng. Fail. Anal.* **2015**, *3*, 68–72. [CrossRef]
12. Karmacharya, S.; Goswami, T. Analysis of a clinically failed, mechanically intact, reconstructive compression plate. Case Studies in Engineering Failure Analysis. *Trauma Cases Rev.* 2015. [CrossRef]
13. Majid, K.; Baker, K.C.; Baker, E.A.; Koueiter, D.M.; Herkowitz, H.N. High Rate of Corrosion of 316L Stainless Steel Plates Used in Posterior Thoracolumbar Fixation: A Retrieval Study. In Proceedings of the 55th Annual Meeting of the Orthopaedic Research Society, Las Vegas, NV, USA, 22–25 February 2011.
14. Varadharajan, S.; Laughlin, R.; Goswami, T. Analysis of clinically failed, mechanically intact, hemi-toe implant. Case Studies in Engineering Failure Analysis. *J. Biomed. Eng. Biosci.* **2016**, *3*, 22–25.
15. Sudhakar, K.V. Metallurgical investigation of a failure in 316L stainless steel orthopaedic implant. *Eng. Fail. Anal.* **2005**, *12*, 249–256. [CrossRef]

16. Sivakumar, M.; Kamachi Mudali, U.; Rajeswari, S. Investigation of failures in stainless steel orthopaedic implant devices: Fatigue failure due to improper fixation of a compression bone plate. *J. Mater. Sci. Lett.* **1994**, *13*, 142–145. [CrossRef]
17. Marcomini, J.B.; Baptista, C.A.R.P.; Pascon, J.P.; Teixeira, R.L.; Reis, F.P. Investigation of a fatigue failure in a stainless steel femoral plate. *J. Mech. Behav. Biomed. Mater.* **2014**, *38*, 52–58. [CrossRef] [PubMed]
18. Cahoon, J.R.; Paxton, H.W. Metallurgical analyses of failed orthopedic implants. *J. Biomed. Mater. Res.* **1968**, *2*, 1–22. [CrossRef] [PubMed]
19. Azevedo, C.R.F. Failure analysis of a commercially pure titanium plate for osteosynthesis. *Eng. Fail. Anal.* **2003**, *10*, 153–164. [CrossRef]
20. De Medeiros, R.C.; de Moura, A.L.; Rodrigues, D.C.; Mendes, M.B.M.; Sawazaki, R.; Moreira, R.W.F. Fractographic analysis of 2.0-mm plates with a screw locking system in simulated fractures of the mandibular body. *J. Oral Maxillofac. Surg.* **2014**, *72*, 1130–1137. [CrossRef] [PubMed]
21. Roffey, P. Case study: Failure of a high nitrogen stainless steel femoral stem. *Eng. Fail. Anal.* **2012**, *20*, 173–182. [CrossRef]
22. Gundapaneni, D.; Slocum, M.; Paul, P.; Christof, S.; Goswami, T. Failure Investigation of a Tibiotalocalcaneal Arthrodesis System. *Eng. Fail. Anal.* **2016**, *59*, 588–604.
23. SYNTHES. 3.5 mm LCP Proximal Humerus Plate—Stainless Steel and Titanium Technique Guide. 2002. Available online: http://synthes.vo.llnwd.net/o16/LLNWMB8/US%20Mobile/Synthes%20North%20America/Product%20Support%20Materials/Technique%20Guides/DSUSTRM10161133_ProxHumPl_STG_150dpi.pdf (accessed on 9 June 2015).
24. SYNTHES. Small Fragment Locking Compression Plate (LCP) System. Stainless Steel and Titanium-Technique Guide. 2002. Available online: http://synthes.vo.llnwd.net/o16/Mobile/Synthes%20North%20America/Product%20Support%20Materials/Technique%20Guides/SUSA/SUTGSmFrgLCPJ3908I.pdf (accessed on 9 June 2015).
25. *Annual Book of ASTM Standards (Vol. 13.01)*, American Society for Testing and Materials: West Conshohocken, PA, USA, 2005.
26. Azevedo, C.; Hippert, E. Failure analysis of surgical implants in Brazil. *Eng. Fail. Anal.* **2002**, *9*, 621–633. [CrossRef]
27. Powell, G.W.; Mahmoud, S.E. *ASM Handbook*, 9th ed.; American Society for Metals: Metals Park, OH, USA, 1986; Volume 11.

metals

MDPI

Article

Failure Analysis of PHILOS Plate Construct Used for Pantalar Arthrodesis Paper II—Screws and FEM Simulations

Farah Hamandi [1], Richard Laughlin [2] and Tarun Goswami [1,*]

[1] Department of Biomedical, Industrial, and Human Factors Engineering, Wright State University, Dayton, OH 45435, USA; hamandi.3@wright.edu

[2] Department of Orthopedic Surgery, Sports Medicine and Rehabilitation, Wright State University, Dayton, OH 45435, USA; rtlaughlin@premierhealth.com

* Correspondence: tarun.goswami@wright.edu; Tel.: +1-937-775-5120

Received: 19 March 2018; Accepted: 16 April 2018; Published: 18 April 2018

Abstract: A fractured stainless steel 3.5 mm proximal humerus internal locking system (PHILOS) plate and screws were investigated in this paper. This plate was used for ankle arthrodesis of a 68-year-old female with a right ankle deformity. Both the plate and screws were considered in this investigation. Optical and scanning electron microscopes (SEM) were used to document fracture surface characteristics, such as extensive scratching, plastic deformation, rubbed surfaces, discoloration, and pitting, along with cleavage, secondary cracking, deposits of debris, striations, and dimples. Indications of these features show that the plate failed by corrosion fatigue, however, overloading separated the screw(s) in two parts. Radiographic evidence shows that the screws failed ahead of the plate from the proximal end. Three-dimensional models of the plate and the screws: cortical, locking, and cannulated, were constructed using Solidworks and imported in ANSYS Workbench 16.2 to simulate the loading conditions and regions of stress development. Statistical analysis was conducted to understand the impact of different factors on the maximum von Mises stresses of the locking compression plate. These factors were the load, screw design pattern, coefficient of friction between the plate and screws, and cortical screw displacement. In summary, the finite element simulation of the plate validates the fractographic examination results. The following observations were made: (a) as the angle between the screws and the plates increased, the von Mises stresses increased in the cortical screws; and (b) the stress in the locking screws was lower than that of the cortical screws, which may be due to locking the screws with fixed angles onto the plate. Finally, fractographic examination of the cortical and locking screws supports the mechanism of corrosion-fatigue fracture from crack initiation sites, pits, due to the presence of inclusion bodies for this material (ASTM standards F138-03 and F139-03) documented for the plate in Paper I.

Keywords: locking plate; fracture; finite element; implant; cortical screw; locking screw

---

1. Introduction

Numerous methods of fixation have been advocated for tibiotalocalcaneal (TTC) arthrodesis, including external fixation, screws [1], intramedullary nails [2], dynamic compression plate fixation [3], and the use of locking plates. Intramedullary nails have several advantages over other internal fixation devices, for instance, better bending and torsion strength [4,5]. However, this type of fixation may not be used with severe-deformity patients [6]. Plates are widely used in bone internal fixation. Different materials are used for medical implants, such as Ti-6Al-4V, 316l stainless steel, Co Cr. 316L stainless steel is used extensively in internal fixation devices due to its excellent properties, such as its competitive price point and compatibility with the human body. On the other hand, corrosion fatigue can drastically affect the lifetime of the internal fixation devices and lead to catastrophic failure.

As screw positioning is one of the main problems that leads to the failure of plates, locking compression plates (LCP) are preferred in arthrodesis. With this technique, the screw head threads are fixed to the plate with a specific angle that helps in reducing the risk of screw failure. Locking screws increase the stabilization of fractures, but they are more expensive than other types of screws. Hybrid fixation uses a combination of both locking and non-locking screws, which improve the stabilization economically. The locking and hybrid construct principles are outlined in Goswami et al. [7].

1.1. Bone Screw Aspects of Design

A screw comprises of a head, shaft, thread, and tip.

- The screw heads can be crossed, hexagonal, or star. The crossed screw head driver has a good advantage in torque transmission [8]. However, a hexagonal driver provides a strong and insensitive alignment connection with the screw.
- The screw shaft represents a smooth link with no threads and might be (1) fully threaded with the cross-sectional diameter decreasing from the head to the bottom tip of the screw; (2) has a similar diameter for both the shaft and the thread root and provides strength and alignment inside the holes, however, it may require over drilling during surgery; and (3) have a similar diameter for both the shaft and the thread major diameter, however, with a weaker shaft [9].
- Screw threads with a constant angle increase the depth and pitch of the screw, while decreasing the pitch of the screw decreases the depth but keeps the angle constant. The surface of the thread should be perpendicular to the pullout load direction [9].
- The screw tip can be cork, blunt, self-tapping, or self-cutting. A self-tapping tip has a sharp flue that needs a lower amount of force. However, it needs 30–40% more torque to be placed [9].

1.2. Biomechanics of Bone Screws

There are many differences among the cortical, cancellous, locking, and cannulated screws. The head of the locking screw has threads. Additionally, the size, core, and pitch of the threads change depending on the type of the screw. The cancellous screw has a larger pitch and a larger core diameter than the cortical screw. In addition, the tip of the screw is either rounded or fluted [10].

When using non-locking screws, a compressive force between the plate and the bone occurs with enough friction to provide the stability with a higher stress than the locking screws. On the other hand, the initial stress develops due to the tightening of the head inside the plate, followed by a compressive force that occurs on the bone around the locking screw. Additionally, the angle does not change between the locking screw and the plate that provides higher resistance against bending [11]. The cannulated screw has less strength than the non-cannulated screw as it has a larger core diameter [12]. The main mechanisms for screw failure are the excessive torque that leads to shear failure or the loading of the plate in a different direction than the screw, which leads to bending failure [9].

Studies showed that 42% of the torque that is applied to the screw is lost to overcoming the friction between the screw and the bone [9,13]. This is one of the main reasons that surgeons apply high force and torque, which leads to higher shear stress on the screw, where the shear stress is directly proportional to the torque and inversely proportional to the diameter. Shear failure force is proportional to the material ultimate shear stress and the thread shear area [14].

In addition, there are two main factors that affect the amount of torque required for the insertion of the screw, which are the tapping and lubrication. Perren [8] showed that 43% of torque loss happens due to the screw-plate friction, 42% of torque loss due to the screw threads-bone friction, and only 15% of torque transforms into an axial force. While, Hughes and Jordan [13] showed that 50% of torque occurred due to the screw-plate and screw-bone friction, 10% of torque loss is due to the screw threads-bone friction, 35% of torque loss is due to cutting threads, and only 5% of torque transforms into an axial force. In contrast, the lubricated and tapped screw friction decreased by 50%, 0% of torque loss is due to the screw threads cutting, and 65% of torque transformed into axial force [9].

There are many designs of plates that can be used for tibiotalocalcaneal fixation or the fixation of a tibial bone fracture. Unfortunately, the lack of resources makes the best fixation option not always available. This issue has resulted in the use of humeral fixation plates for tibiotalocalcaneal arthrodesis. There are several studies that focused on the capability of using the humeral locking plate for tibiotalocalcaneal arthrodesis, as shown in Table 1, wherein 2007, a study on 18 patients showed that 94.4% of the patients had successful fusion after approximately 20 weeks when the proximal humeral locking plates were used for obtaining tibiotalocalcaneal arthrodesis [1]. Additionally, another study in 2011 reported successful results when the same plate was used for ankle arthrodesis [15]. In addition, in 2016, Shearman et al. showed that 81% of patients had satisfactory results when proximal locking plates were used for obtaining tibiotalocalcaneal arthrodesis [4].

Table 1. Studies focused on the capability of proximal humerus internal locking system (PHILOS) for tibiotalocalcaneal arthrodesis.

Authors	Year	Title	Age	Gender Male:Female	Details
Ahmad, J., Pour, A.E., and Raikin, S.M. [1]	2007	The modified use of a proximal humeral locking plate for tibiotalocalcaneal arthrodesis	22–72 years Mean: 54.2	11:6	18 patients showed that 94.4% (17 of 18) of the patients had successful fusion after approximately 20 weeks when the proximal humeral locking plates were used for obtaining tibiotalocalcaneal arthrodesis.
Shi, Z., Zhang, C., Gu, W., Zhang, C., and Zeng, B. [15]	2011	Ankle arthrodesis by lateral malleolus osteotomy and internal fixation with locking proximal humeral plate	36–67 years Mean: 48	10:8	Reported successful results when the same plate was used for ankle arthrodesis.
Haider Twaij and Dev Damany	2013	PHILOS humerus plate for a distal tibial fracture	51	1:0	A successful result for the case.
Shearman, A.D., Eleftheriou, K.I., Patel, A., Pradhan, R., and Rosenfeld, P.F. [4]	2016	Use of a Proximal Humeral Locking Plate for Complex Ankle and Hindfoot Fusion	25–74 years Mean: 56.1	11:10	17 of 21 patients (81%) had satisfactory results when proximal locking plates were used for obtaining tibiotalocalcaneal arthrodesis.

In the present study, a failed PHILOS construct was examined and a computational simulation performed to validate the regions of stress development causing the physical failure of the system.

2. Background

A fractured stainless steel 3.5 mm LCP proximal humerus locking plate (PHILOS), originally manufactured by Synthes, is investigated for the failure mechanism(s) responsible for its failure. A typical LCP proximal humerus locking plate is shown in Figure 1. It can be seen that the plate was inverted to handle the anatomical shape of the tibiotalocalcaneal area, where the proximal part of the plate was fixed to the calcaneus and talus. The fractured pieces are analyzed to determine the mechanism of failure. In Paper I, the PHILOS plate failure analysis was carried out and a state of the art literature review of hybrid constructs is summarized [16]. This paper not only presents the results of finite element simulations of the PHILOS construct (plate and screws) but also examines the failure mechanisms of the screws that failed.

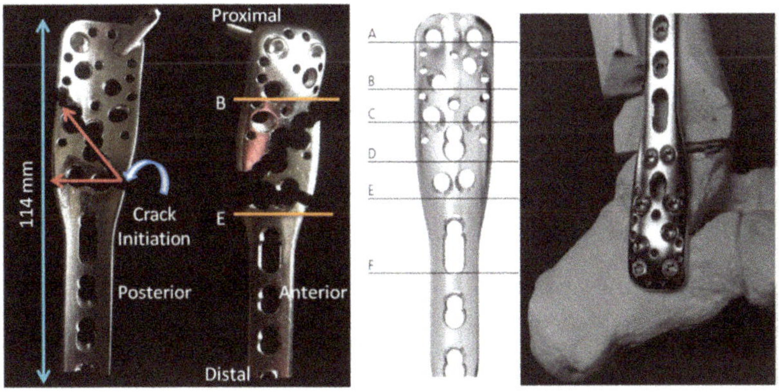

Figure 1. The fractured 3.5 mm LCP proximal humerus locking plate shaft showing the crack initiation and propagation in two directions (red arrows) (**left**). The proximal part of the plate is divided into six levels, from A to F (**middle**). It can be seen that the plate is inverted to align with the anatomical shape of the tibiotalocalcaneal area, where the proximal part of the plate was fixed to the calcaneus and talus (**right**).

The PHILOS plate and screws were used for a 68-year-old patient that had a right ankle plantar arthrodesis, deformation on the right ankle, and dislocation on the lateral subtalar joint. The distal part of the plate was fixed to the lateral part of tibia and calcaneus by 3.5 mm cortical screws. Additionally, 3.5 mm locking screws were used to fix the proximal part of the plate. Furthermore, a 6.5 mm cannulated screw was used to apply compression on tibiotalar and subtalar joints. In addition, 3.5 mm screws with washers were used to support the fusion of the talonavicular joint. After six years, the plate and the screws were removed from the body. Table 2 summarizes all the parts that are submitted for investigation.

Table 2. Pieces of the PHILOS system submitted for investigation.

Name of Device	Size	Company	Serial No.	Date of Insertion	Failure First Noticed
LCP Proximal Humerus Locking Plate	3.5 mm	DePuy Synthes	N/A	08/2008	2014
Cortical Screws	3.5 mm	DePuy Synthes	N/A	08/2008	2010–2014
Locking Screws	3.5 mm	DePuy Synthes	N/A	08/2008	2011–2014
Cannulated Screw	6.5 mm	Zimmer	N/A	08/2008	2013

Note: X-ray images with respect to dates are shown in [16].

3. Visual Examination

3.1. The Plate

Visual examination demonstrated that the LCP plate was fractured into three pieces. The anterior and posterior views of the plate are shown in Figure 1. The proximal part of the plate can be divided into six levels, from A to F [17], as shown in Figure 1.

The plate failed at levels B–E. Additionally, Figure 1 shows that the crack initiated from under one of the locking holes near the middle of the plate at level (E). This fracture progressed in two directions (the red arrows), one in the diagonal direction and another in the perpendicular direction, which caused the plate to fail into three pieces. The distal part of the plate shaft had been removed to perform some additional tests. Observing the plate under the optical microscopy showed a considerable amount of scratches and pitting. Details of the plate failure have been discussed in Paper I [16].

3.2. The Screws

Twelve screws were submitted for examination, as shown Figure 2. Four were 3.5 mm cortical non-locking screws, and three (75%) of these screws fractured into two pieces. The X-ray images showed that the first cortical screw failed after two years of implantation [16] and the other two screws failed after five to six years [16], visible in X-rays. One of the cortical screw heads remained on the plate after removal. Seven 3.5 mm locking screws were submitted for investigation, and three (42.8%) of these screws fractured into two pieces. It was difficult to visualize the time of failure of the locking screws after the implantation in the X-ray images because the screw density was high at the area of insertion of the locking screws. One of the locking screw heads and other undamaged locking screws remained in the plate after removal. In addition, one 6.5 mm cannulated screw fractured within five years of implantation, as was observed in the X-ray images [16]. No physical damage to the devices that may have led to the failure was observed. Radiographic evidence, as seen in Figure 1, shows that the screws failed ahead of the plate from the proximal end.

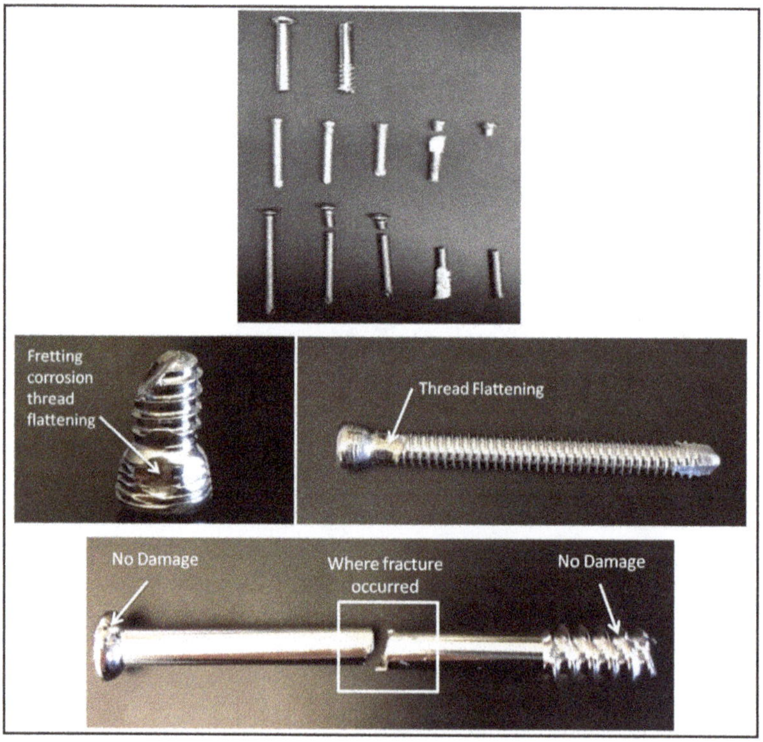

Figure 2. Screw visual observation showed the fractured screws (**upper**), some thread flattening (**middle**), and the fractured 6.5 mm cannulated screw (**below**).

4. Fractographic Examination

Optical and scanning electron microscopes were used to document the fracture surface characteristics such as extensive scratching, plastic deformation, rubbed surfaces, discoloration, and pitting, along with cleavage, cracking, deposits of debris, striations, and dimples. A PEMTRON scanning electron microscope (SEM PS-230, Seoul, Korea) with 10 KV was used to perform the fractographic examination. Fractographic analysis was performed on the plate and four screws in the fractured regions. By examining the fractured surface of the plate, quasi-cleavage fracture was observed, as shown in Appendix A (Figure A1). The literature showed similar observations that the corrosion–fatigue mechanism was responsible for the failure of the plates [18–20]. However, another study showed that the failure of plates could be a result of improper material [21].

The cortical screws were marked as CS1, CS2, CS3, and CS4 from the proximal to the distal direction. The locking screws were marked depending on the level of insertion, shown in Figure 1, as LA, LB, LC, LD, and LE. Cortical screws CS1 and CS4 were examined with the SEM. Appendix A (Figures A2–A4) show the main details in CS1, where the presence of striations and microcracks can be noticed, while Appendix A (Figures A5–A7) shows the main details in CS4, where the presence of striations in different directions can be noticed. In addition, two locking screws were examined with the SEM. Appendix A (Figures A8–A11) shows the main details for the locking screws.

It can be noticed that the fractured surfaces of the screws were rubbed. Rubbing happened inside the body as the devices failed after two years of insertion and were kept inside the body for four more years, which caused the loss of some of the important features that caused the crack initiation. Additionally, it was observed that the cracks in the fatigue areas of the plate initiated from corrosion pits. In general, the investigation showed that the failure was a result of conjoint bending/torsion loading and corrosion–fatigue cracking that propagated from the bottom of pit(s). The pitting susceptibility of this material is widely documented in the literature arising from inclusions, which were reported in Paper I [16]. The mechanisms governing failure of the screws are similar to that found for the plate [16].

5. Material Conformity

The metallographic qualitative analysis was performed based on ASTM standards F138-03 and F139-03 [22]. A sample from the locking screw was taken. Cutting, mounting, and polishing were performed to prepare the sample. The polishing was done with different grades of silicon carbide papers (320–600 grit) and diamond abrasive solutions (9–0.01 μm). After the polishing, X-ray energy dispersive spectroscopy (EDS, Thermo Fisher Scientific, Hillsboro, OR, USA) was performed on the screw sample surface, and the results were compared with the ASTM standards F138-03 and F139-03 reported for stainless steel 316L to ensure that there is no significant difference that might have caused the failure [18,23]. Figure 3 shows the peaks of different element weight percent in the screw, and the data on each area is summarized in Table 3. Additionally, ASTM standard F139-03 has set additional requirements to ensure that the material meets the specifications, which is shown in Equation (1):

$$\%\,\mathrm{Cr} + (3.3\ \times \%\mathrm{Mo})\ \geq 26.0 \qquad (1)$$

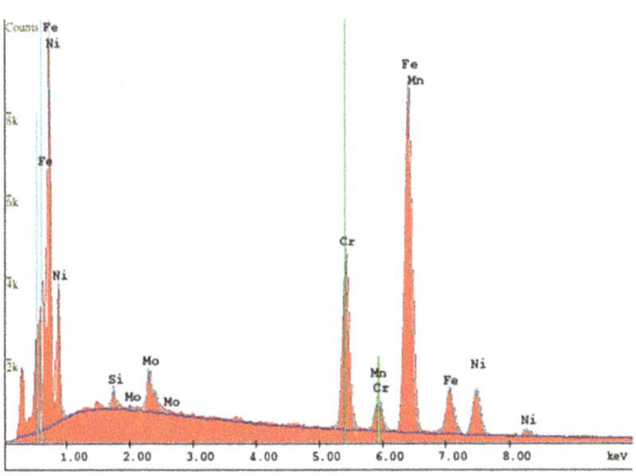

Figure 3. Energy dispersive X-ray analysis shows the peaks of different element weight percent in the cortical screw sample.

Table 3. X-ray energy dispersive spectroscopy (EDS).

Composition	ASTM Standards F138-03 and F139-03	Screw			
	(Min–Max)	Area 1	Area 2	Area 3	Average
%Mn	0–2.04	1.63	1.51	1.62	1.57
%Si	0–0.80	0.62	0.63	0.42	0.56
%Cr	16.80–19.20	18.57	18.68	18.71	18.65
%Mo	2.15–3.10	2.97	3.09	2.36	2.81
%Ni	12.85–15.15	14.83	14.94	14.63	14.8

The screw confirmed to the ASTM standard and there is no significant difference that might have caused the failure.

6. Computational Simulations of Failures

6.1. Finite Element Analysis

Three-dimensional models of the plate and the screws: cortical, locking, and cannulated were constructed using SolidWorks (Dassault Systèmes SolidWorks Corp., Concord, MA, USA) and imported in ANSYS Workbench 16.2 (ANSYS Inc., Canonsburg, PA, USA) to simulate the loading conditions and regions of stress development. The dimensions of the cortical, locking, and cannulated screws were taken from the Synthes brochure [24], and the PHILOS plate was measured from the submitted samples, Figure 4. Models were generated depending on patient-specific geometric information that was obtained manually from the X-ray images. The modeling of each screw has depended on the angle of its insertion inside the bone of this patient. However, locking screw placement angles were not clearly visible in the X-ray due to the presence of too many screws that were used in the fixation, Figure 5. That being said, the locking screws placement angles were measured manually by inserting locking screws inside a sample plate. Figure 6 shows the angles of the locking screws at level A. Locking screws at level A were inserted slightly upward with a 30° angle above the neutral axis of the center of the screw hole and parallel to each other. The locking screws at level B were inserted perpendicular to the plate with an inward 30° angle. The locking screws at level C were inserted upward with a 40° angle above neutral axis and outward 30° angle. The locking screws at level D were inserted slightly upward with a 10° angle above the neutral axis. Finally, the locking screw at level E was inserted upward with a 30° angle above the neutral axis and slightly outward. The heads of the locking screws were fixed with no displacement allowed. Additionally, the material properties of stainless steel 316L were assigned according to the ASTM standard [22], shown in Table 4. The cortical and cancellous components were modeled with a 12.7 GPa and a 0.9 GPa modulus of elasticity and a 0.3 and a 0.2 Poisson's ratio, respectively [25]. Materials were treated as homogeneous, isotropic, and a linear elastic analysis was carried out. All the results were within the limitation of this assumption assuming all loads transferred from bone to the plate. Figure 7 shows the loading conditions that have been applied to the plate and the screws.

It is claimed that when a PHILOS plate is closed with all the screw holes, the fixation is stable and fracture union takes place with time. However, with the current subject, the device started to fail before the bone union took place, which caused some displacement and led the screws to fail after two years, visible when X-rays were taken. We ignored the effect of screw failure on the adjacent level screws both in terms of construct plate-screw placement angles and load transfer. As there is no viable way to identify the amount of screw micro-motion in this construct, it was assumed that displacement in the cortical screws within the range of 0.05–0.25 mm occurring within the few months after load bearing based on our research on hybrid constructs where the stiffness drop occurs within the first 5000 fatigue cycles [7] resulting in altering the coefficients of friction between the screws and bone; 0.5 with all screws intact to 0.1 when screws began to fail. All the locking screws were assumed to be fixed, while some displacement was allowed for the cortical screws along the x-axis, as shown in Figure 8, simulating human gait to reflect loading and deformation. The loading was applied in two steps; (1) the preload torque for the installation of the screw, and (2) the axial loads. The applied load on the ankle joint might reach 3.5 times of the body weight in the gait cycle during the push-off stage [26]. The body weight of the patient was assumed to be 80 kg in this study. However, for patients with abnormal gait cycles, the applied load may reach four times of the body weight (3000 N) or higher. To simulate the contact analysis in the model, the target and contact surfaces between the plate and bone, the plate and screws, and bone and screws were defined by not merging the nodes between the components and assuring the union and the transformation of the loads and the deformation between the plate, screws, and bone with a coefficient of friction of 0.4 between the plate and the bone [27], 0.35 between the plate and the screws [28], and five different coefficients of friction were obtained between the screws and the bone (0.1, 0.2, 0.3, 0.4, and 0.5) [27]. Furthermore, five different axial loads were tested to observe the PHILOS device in normal and extreme situations with axial loads ranging from 500–2000 N with a torque load 5.0 Nm [29,30]. The loading in the screws had been inclined to match with the insertion angle of each of the screws. In addition, two screw pattern designs (Figure 8) were assumed in order to investigate the effect of screw position on the maximum stress of the plate: design (A) where the screws inserted at levels A, C, D, and E, and design (B) where the screws inserted at levels A, B, C, and E, as shown in Figure 8, and the loading and boundary conditions are shown in Figure 7. In addition, the tetrahedral element type was used in this study. The FE meshing was automatically adapted through Ansys, where convergence tools were used with 5% convergence, which resulted in approximately 50,000 elements and 85,000 nodes, as shown in Figure 7. Some studies suggested removing of unused screw holes from the model, however, in this study all the holes were included in the design, and only filled with screws per designs A and B, Figure 8.

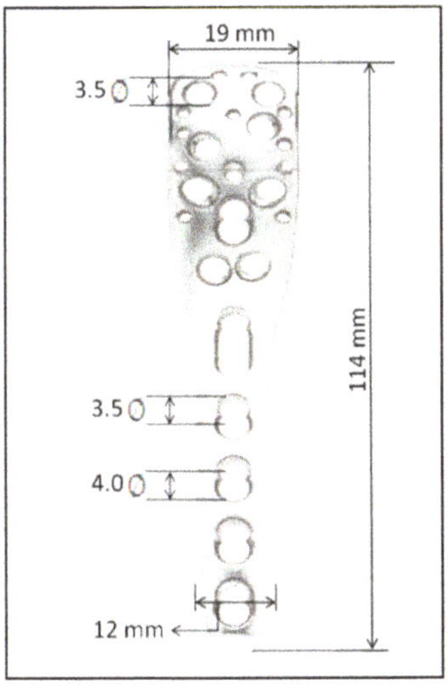

Figure 4. The dimensions of the 3.5 mm standard locking compression plates (LCP) proximal humerus locking plate.

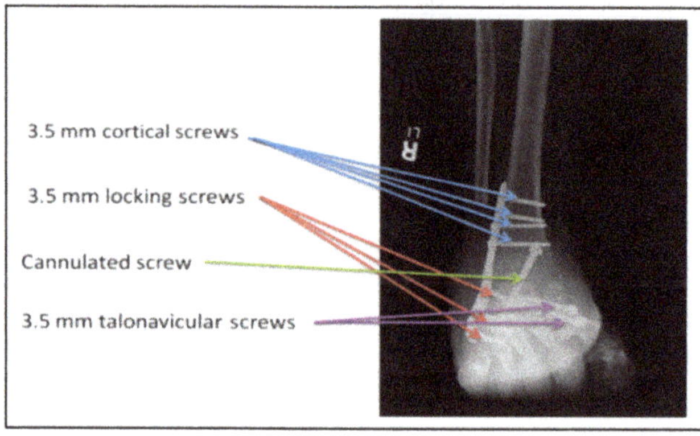

Figure 5. X-ray image showing the screws used in the tibiotalocalcaneal fixation.

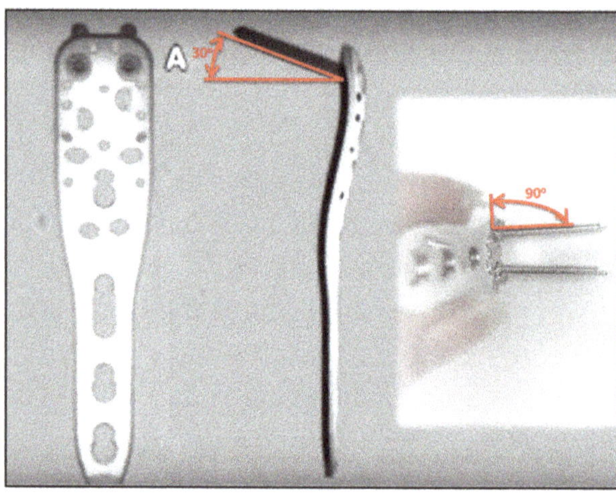

Figure 6. Variable angle placement of the locking screws at level A.

Table 4. Stainless Steel 316L material properties [22].

Material Properties	316L Stainless Steel
Elastic Modulus (GPa)	200
Shear Modulus (GPa)	82
Poisson's Ratio	0.265
Mass Density (g/m^3)	8.027
Tensile Strength (N/mm^2)	860
Yield Strength (N/mm^2)	690

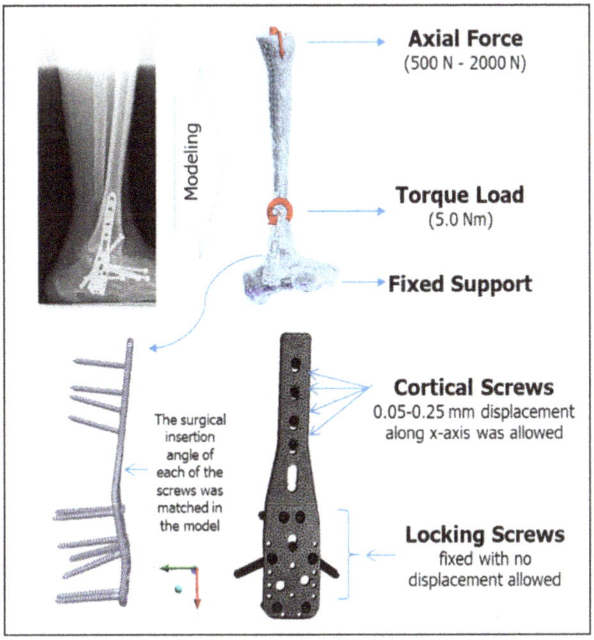

Figure 7. The loading and boundary conditions applied on the model.

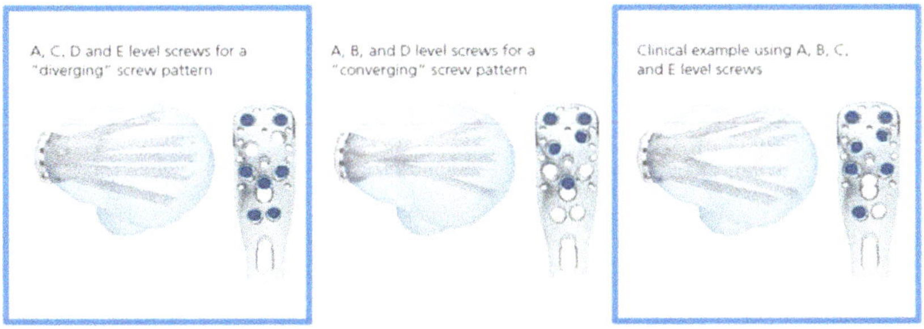

Figure 8. Screw pattern designs. Design A (**left**) and design B (**right**) [17].

6.2. The Results

The maximum von Mises stress in the plate was obtained with two screw pattern designs, five axial loads and coefficients of friction between the screw and the plate, and five cortical screw displacements, and the results are summarized in Table 5. Figure 9 shows the von Mises stresses of the 316L SS plate with screw design B, 500 N axial load, 0.5 coefficient of friction between the screw and the plate, and 0.1 mm cortical screw displacement. It was observed that the maximum stresses are in the levels B–E, which validates the visual and fractography examinations. Additionally, the maximum stresses were between 201 and 346 MPa for 500–750 N axial loading, which is significantly lower than the yield strength (σ_y = 690 MPa) of the material. However, for the fatigue limit at 10^7 cycles, the maximum stresses should be equal to, or less than, 440 MPa to be in the safe region [31]. On the other hand, the maximum stresses were between 504 and 692 MPa for 1500–2000 N axial loading, which is significantly higher than the failure criteria used. At this level, the developed stresses are likely to cause plastic deformation and failure prematurely. If the environmental effects were included, the corrosion fatigue crack propagation may activate at this level of loading. Since the ankle deformity gives rise to the increased level of force development in the range of 4–8 times the body weight, the extrapolated von Mises stress may reach as high as 900 MPa. In addition, the simulations show that higher coefficients of friction reduce the von Mises stress for both pattern designs and for different loading conditions. As a result estimation of the actual loading behavior of the plate from the activities of daily living of the individual, together with mechanical parameters such as the coefficient of friction, loosening, and axial loading in the screw and plate, will be very difficult to isolate and predict the combined mechanisms and failure behavior.

Table 5. The maximum von Mises stresses of the plates were obtained for five different loads, five coefficients of friction, five cortical screws displacement, and two screw pattern designs.

Axial Force (N)	Coefficient of Friction	Design A					Design B				
		Non-Locking Screws Displacement (mm)									
		0.05	0.10	0.15	0.20	0.25	0.05	0.10	0.15	0.20	0.25
500	0.10	219.81	221.90	224.19	226.72	229.49	220.86	222.95	225.24	227.76	230.54
	0.20	210.54	211.70	212.97	214.38	215.92	211.59	212.75	214.02	215.42	216.97
	0.30	205.39	206.03	206.74	207.52	208.38	206.44	207.08	207.79	208.57	209.43
	0.40	202.53	202.88	203.28	203.71	204.19	203.58	203.93	204.33	204.76	205.24
	0.50	200.93	201.13	201.35	201.59	201.86	201.99	202.19	202.40	202.64	202.91
750	0.10	330.26	333.39	336.83	340.62	344.78	331.29	334.42	337.86	341.65	345.81
	0.20	316.35	318.09	320.00	322.10	324.42	317.38	319.12	321.03	323.14	325.45
	0.30	308.62	309.59	310.65	311.82	313.11	309.66	310.62	311.69	312.85	314.14
	0.40	304.33	304.87	305.46	306.11	306.82	305.37	305.90	306.49	307.14	307.85
	0.50	301.95	302.24	302.57	302.93	303.33	302.98	303.28	303.61	303.97	304.36
1000	0.10	440.70	444.87	449.46	454.50	460.06	441.72	445.89	450.48	455.53	461.08
	0.20	422.15	424.47	427.02	429.82	432.91	423.18	425.49	428.04	430.85	433.93
	0.30	411.85	413.14	414.56	416.11	417.83	412.88	414.16	415.58	417.14	418.85
	0.40	406.13	406.84	407.63	408.50	409.45	407.15	407.87	408.66	409.52	410.47
	0.50	402.95	403.35	403.78	404.26	404.79	403.97	404.37	404.81	405.29	405.82
1500	0.10	551.13	556.34	562.08	568.39	575.33	552.15	557.36	563.10	569.41	576.35
	0.20	527.95	530.85	534.04	537.54	541.40	528.97	531.87	535.06	538.56	542.42
	0.30	515.08	516.68	518.46	520.40	522.55	516.10	517.70	519.48	521.42	523.56
	0.40	507.92	508.82	509.80	510.88	512.07	508.94	509.84	510.82	511.90	513.09
	0.50	503.95	504.44	504.99	505.59	506.25	504.97	505.46	506.01	506.61	507.27
2000	0.10	661.56	667.82	674.70	682.28	690.61	662.58	668.84	675.72	683.29	691.62
	0.20	633.75	637.23	641.05	645.26	649.88	634.77	638.24	642.07	646.27	650.90
	0.30	618.30	620.23	622.35	624.69	627.26	619.31	621.25	623.37	625.71	628.28
	0.40	609.71	610.79	611.97	613.27	614.69	610.73	611.80	612.98	614.28	615.71
	0.50	604.94	605.54	606.20	606.92	607.71	605.96	606.56	607.21	607.93	608.73

On the other hand, the maximum von Mises stresses increased with the increase in cortical screw displacements, as the screws began to fail, as shown in Figures 10 and 11. The maximum von Mises stresses of the cortical screws occurred near the head of the screw when the screw was perpendicular to the plate, as shown in Figure 12 (will be further discussed in Section 7). However, the maximum von Mises stress of the cortical screw was near the middle of the screw when the screw was at an upward angle, which will be further discussed in Section 7. In addition, the maximum von Mises stresses of the locking screws were distributed across the screws and were higher at levels A, C, and E, where the screws have slightly upward angles. Figure 13 shows the von Mises stress of the locking screw at level C and the screw insertion angles. Table 6 shows the maximum von Mises stresses of the cortical and locking screws. Finally, the von Mises stress of the cannulated screw is shown in Figure 14, and it is observed that the stresses are higher in the shaft of the cannulated screw.

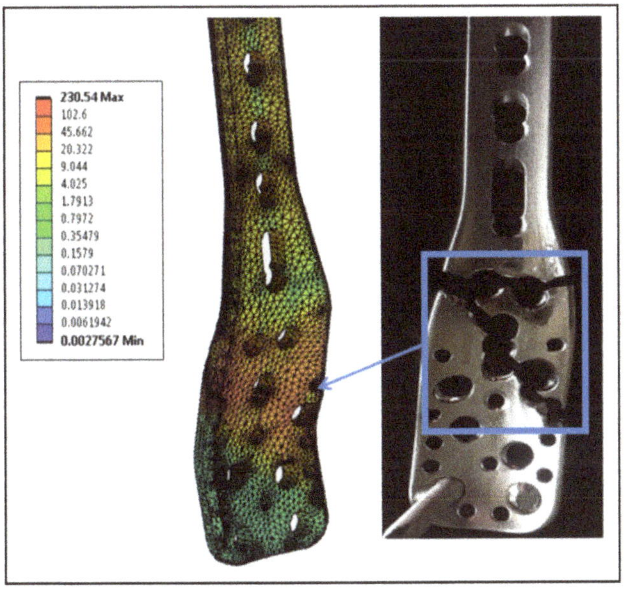

Figure 9. The von Mises stresses in the 316L SS plate with screw design B, 500 N axial load, 0.5 coefficient of friction between the screw and the plate, and 0.1 mm cortical screw displacement.

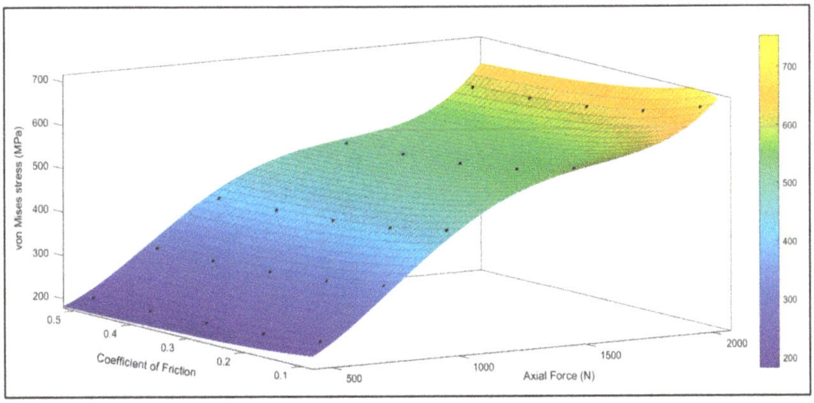

Figure 10. The surface plot of the maximum von Mises stresses in the 316L SS plate with screw design B and 0.25 mm cortical screw displacement vs. the coefficient of friction (COF) and the axial force.

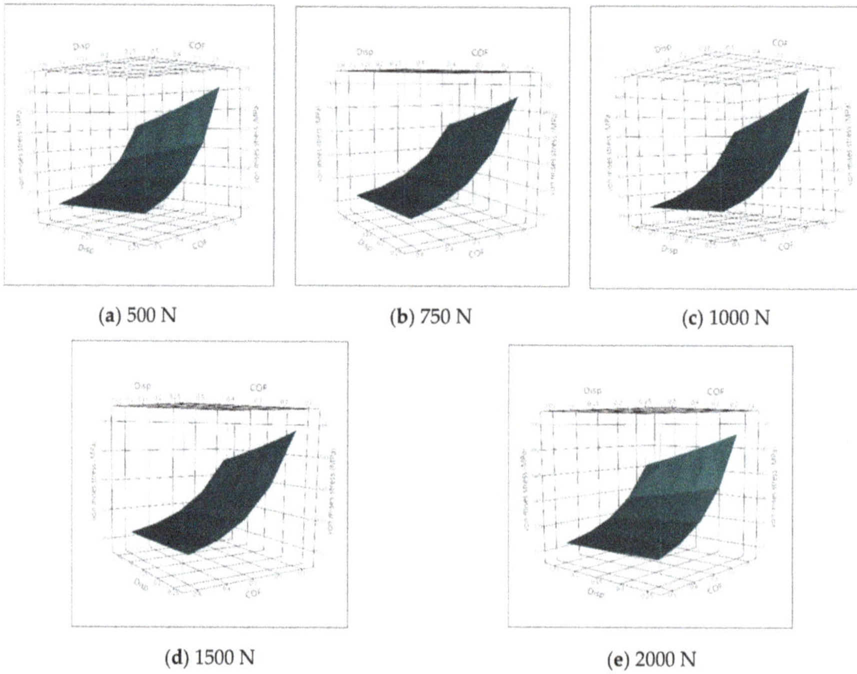

Figure 11. The surface plot of maximum von Mises stresses in the 316L SS plate vs. the coefficient of friction (COF) and the cortical screw displacement (mm) for different axial loads (**a**) 500 N, (**b**) 750 N, (**c**) 1000N, (**d**) 1500 N, (**e**) 2000 N.

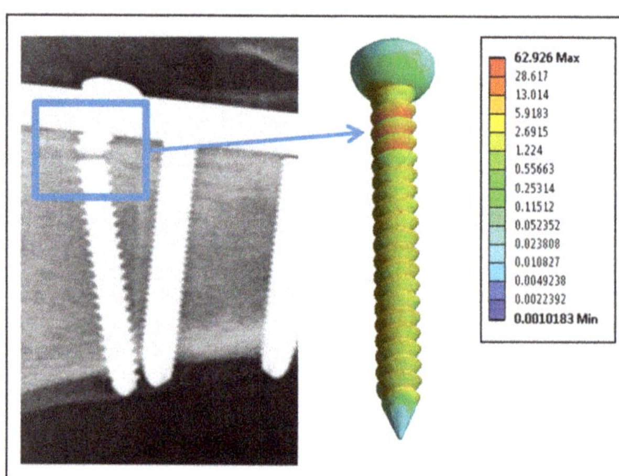

Figure 12. Comparing a two-year post operation X-ray image with the von-Mises stress of the 316L SS cortical screw (CS3).

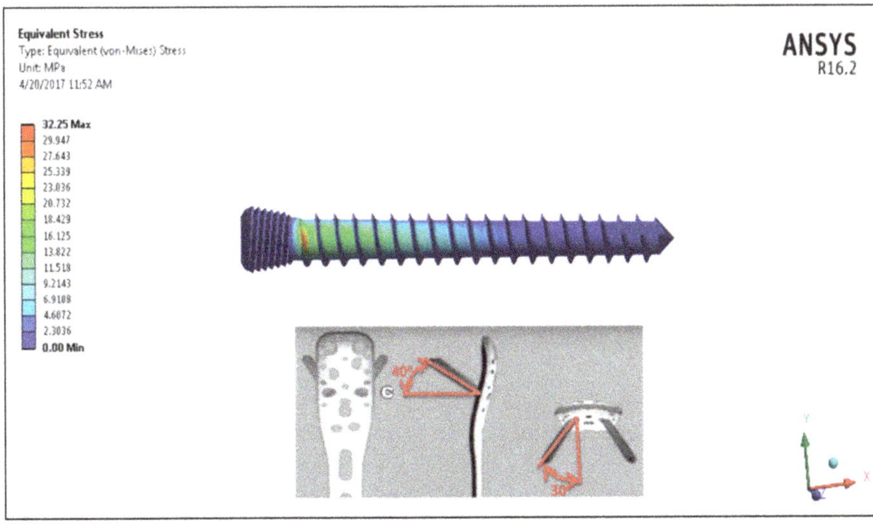

Figure 13. The von Mises stress of the 316L SS locking screw at level C and the screw insertion angles.

Table 6. The maximum von Mises stresses in the screws.

Description	Cortical Screws				Locking Screws				
The Screw	CS1	CS2	CS3	CS4	A	B	C	D	E
Maximum Stress (MPa)	54.164	56.6316	62.926	59.627	31.138	26.4	32.25	28.04	31.9

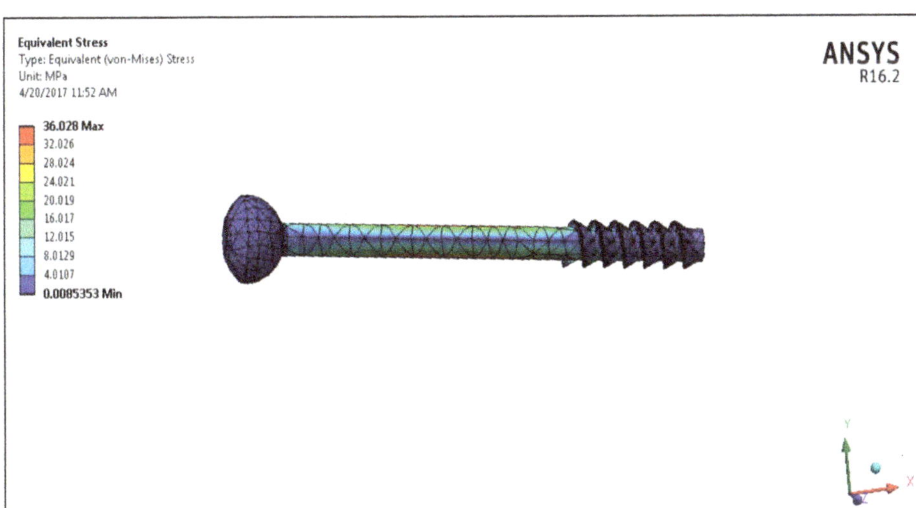

Figure 14. The von-Mises stresses in the cannulated screw.

7. Discussion

This paper not only examines the PHILOS plate computational simulations, but also examines the failure mechanisms of all the screws that failed. On visual examination, it can be noticed that the crack initiated from under one of the locking holes near the middle of the plate. Observing the plate under optical microscopy indicated that the surface of the plate had a considerable amount of scratches and pitting. Seventy-five percent of the cortical screws and 42.8% of the locking screws fractured into two pieces. Chemical, qualitative analysis of our results shows that the screw confirmed to the ASTM standards F138-03 and F139-03 and there is no significant difference that may have caused the failure.

Optical and scanning electron microscopes were used to document the fracture surface characteristics, such as extensive scratching, plastic deformation, rubbed surfaces, discoloration, and pitting, along with cleavage, cracking, deposits of debris, striations, and dimples. Fractographs show a quasi-cleavage fracture. Azevedo [19], Kanchanomai et al. [20], and Majid et al. [32] showed similar observations that the corrosion–fatigue mechanism was responsible for the failure of plates. However, Sivakumar et al. [33] showed that the failure of plates could be a result of improper fixation. On the fracture surface of the cortical screws, we noticed the presence of striations, microcracks, and rubbed surfaces. The presence of the inclusions, as presented in paper I [16] for this material, increased the pitting susceptibility of SS316L and possible transitioning to corrosion fatigue cracking. FEM analysis clearly shows that at any level of load bear, the developed stresses will cause plastic deformation at the sites of screw holes and this plasticity will initiate a crack. Thapa et al. showed a similar fractographic examination for the locking plate [34]. Indications of these features show that the plate failed by corrosion fatigue. However, overloading separated the screws into two parts. Radiographic evidence shows that the screws failed ahead of the plate from the proximal end. Finally, fractographic examination of the cortical and locking screws supports the mechanism of corrosion–fatigue fracture from crack initiation sites due to the presence of inclusion bodies and pits.

In addition, two screw pattern designs were assumed in order to investigate the effect of the screw position on the maximum stress of the plate. Some studies suggested the removal of unused screw holes from the model, however, in this study all the holes used in the simulation provided higher fixation strength and stability. The results of the computational simulation showed that the maximum stresses simulated the loading conditions and regions of stress development. The simulation showed that having higher coefficients of friction reduces the von Mises stresses for both pattern designs and for different loading conditions. On the other hand, the maximum von Mises stresses of the PHILOS plate increased with the increase in cortical screw displacements. The results for the visual, fractography, and quantitative examinations of the cortical, locking, and cannulated screws are summarized in Table 7.

Table 7. Visual, fractography, and quantitative examinations of the cortical, locking, and cannulated screws.

	Screw	Visual Examination	Fractographic Examination	Quantitative Examination
Cortical Screws 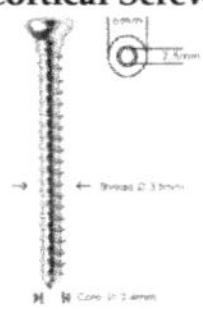	CS1	Plastic deformation, failed into two pieces X-ray image shows it was failed after two years of insertion	Fatigue striations Microcracks A rubbed surface	The maximum von Mises stresses were near the head of the screw
	CS2	No noticeable failure	N/A	The maximum von Mises stresses were near the head of the screw
	CS3	Plastic deformation, failed into two pieces X-ray image shows it was failed after five years of insertion	N/A	The maximum von Mises stresses s were near the middle of the screw
	CS4	Plastic deformation, failed into two pieces X-ray image shows it was failed after six years of insertion	Inclusion particles displaying ductile fracture Fatigue striations	The maximum von Mises stresses s were near the middle of the screw
Locking Screws 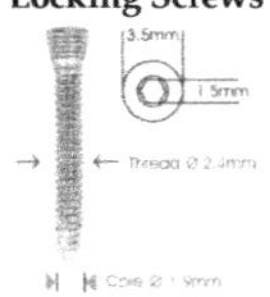	LA	Inserted slightly upward with 30° angle and parallel to the other screw. One screw failed and fractured into two pieces	N/A	The von Mises stresses distributed between 20–40% the screw length Higher than levels b and d
	LB	Inserted perpendicular to the plate and inward 30° angle. No noticeable failure	N/A	The von Mises stresses distributed near the head of the screw Lower than levels a, c, and e
	LC	Inserted upward with a 40° angle and outward 30° angle. One screw failed and fractured into two pieces	Inclusions Fatigue striations	The von Mises stresses distributed between 20–40% the screw length Higher than levels b and d
	LD	Inserted slightly upward with a10° angle. No noticeable failure	N/A	The von Mises stresses distributed near the head of the screw Lower than levels a, c, and e
	LE	Inserted upward with a 30° angle and slightly outward. One screw failed and fractured into two pieces	Inclusions Fatigue striations	The von Mises stresses distributed between 20–40% the screw length Higher than levels b and d
Cannulated Screw 		Plastic deformation, failed into two pieces X-ray image shows it was failed after six years of insertion	N/A	The stresses were higher in the shaft of the screw

The maximum von Mises stresses of the cortical screws were near the head of the screw when the screw was perpendicular to the plate and near the middle when the screws were at upward angles. If we compare the von Mises stresses of the four cortical screws across the cortical screws length, it was observed that the maximum stresses are almost similar for all the screws, but the position of the maximum stresses varied depending on the angle of the screws, as shown in Figure 15. Additionally, the graph shows that the stresses are, at a minimum, at the head of the C1 and C2 screws. Then, a sharp increase in the stresses can be noticed at the shaft of the screw near the head, as this is the area where the screw cuts through the cortical bone and is exposed to the axial load of the bone. Then the stress distribution decreased gradually. On the other hand, C3 and C4 cortical screws were at an upward angle. This angulation increased the stress by 165% on the shaft of the screw than the head of the screw. In the case of locking screws, the maximum von Mises stresses were distributed across the screws. However, the maximum von Mises stresses were only 110% higher at levels A, C, and E where the screws have slightly upward angles. Figure 16 shows a comparison between the von Mises stresses across the locking screw length. It can be noticed from the graph that the maximum stresses were near the heads of the screws and as the screws entering the cortical bone. Finally, it was observed that the stresses were 170% higher in the shaft of the cannulated screw than the head and the tip of the screw. In general, the maximum von Mises stresses experienced by the screws were lower than the fatigue stress limit. However, corrosion in this case may drastically affect the lifetime of the screws by reducing the number of cycles to failure. The numerical analysis was conducted to understand the impact of different factors on the maximum von Mises stresses of the locking compression plate. The statistical analysis showed that the applied load has a significant effect on the maximum stresses, but the screw pattern design does not have a significant effect. In summary, the fractographic examination of the cortical and locking screws supports the mechanism of corrosion–fatigue fracture from crack initiation sites due to the presence of pits and/or high plasticity regions. Additionally, the cracking followed by the rotating bending mechanisms since the axial load and torsion control that behavior. This study validates the physical failure with the computational simulations.

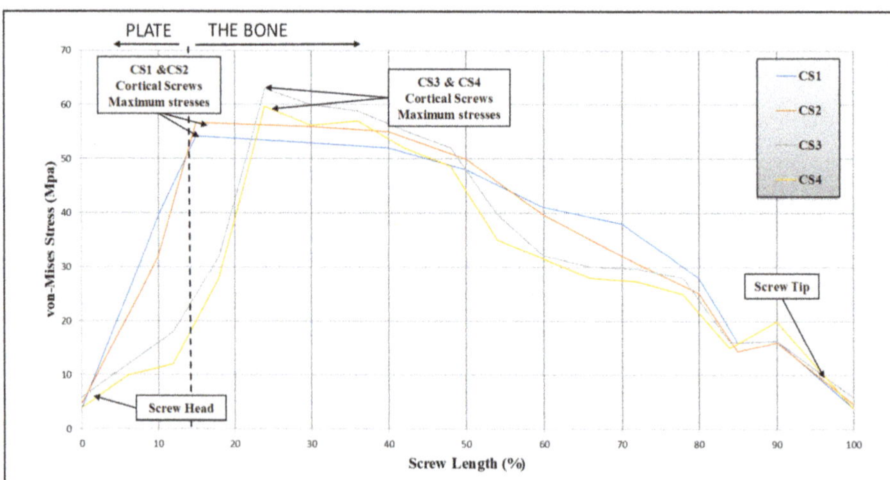

Figure 15. The von-Mises stresses of the 316L SS cortical screws across the screw length.

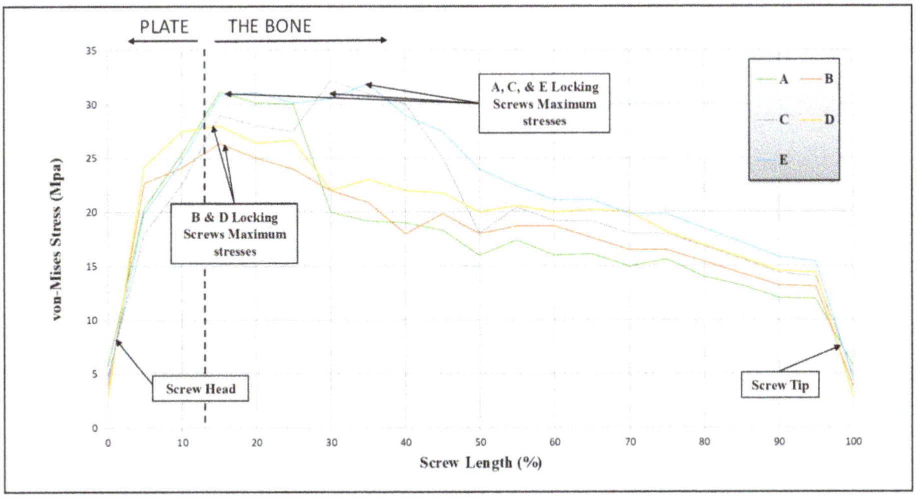

Figure 16. The von-Mises stresses of the 316L SS locking screws across the screw length.

8. Conclusions

Attempts were made to perform qualitative and quantitative failure analyses of the PHILOS construct in this paper. This study clarifies that there are several factors that, together, caused the mechanical and clinical failures of the system summarized below.

1. Fractographic examination of the cortical and locking screws supports the mechanism of corrosion-fatigue fracture from crack initiation sites due to the presence of inclusion bodies and pits and regions of high plastic strains due to load bear. The regular features, such as beach marks and striations, were present indicating fatigue and/or corrosion fatigue controlled the failure process.
2. The simulations showed that the maximum von Mises stresses of the PHILOS plate increased by 2.2% with the increase in cortical screw displacement as expected as the screws began to fail, giving rise to higher macro-motion of the construct at the same time lowering the coefficient of friction between the contacts.
3. The analysis of cortical screws showed that the stresses increased by 7.4% as the angle between the screw and the plate increased.
4. The stresses in the locking screws were lower than the cortical screws by 25.5%, and this may be a result of the locking screws' fixed angles with less range of motion.

The finite element simulation of the plate validated the loading conditions and regions of stress development that supported the visual and fractographic examinations causing the physical failure of the PHILOS system.

Acknowledgments: Allen Jackson, Wright State University, Dayton, Ohio provided access to the SEM facility.

Author Contributions: Farah Hamandi performed the SEM analysis and texture related research, mechanical tests, and the theoretical aspects of the paper. Richard Laughlin performed the revision surgery and removed the device components. Tarun Goswami, led this research study, recruited students, verified the analytical methods and edited or written the body of the paper.

Conflicts of Interest: The authors declare no conflict of interest.

Appendix A

Fractographic Examination

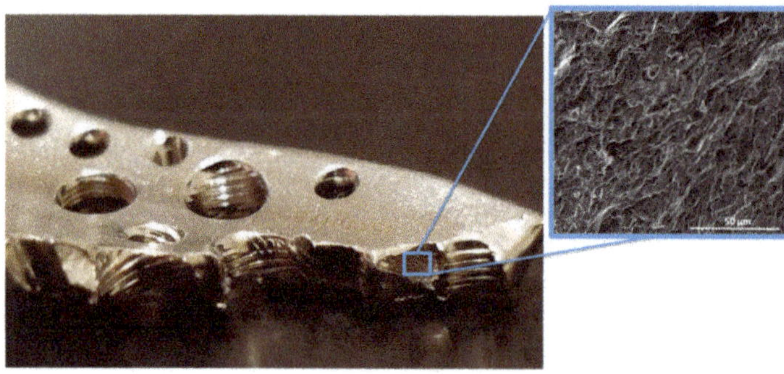

Figure A1. The quasi-cleavages, which are characteristics of fatigue failure (SEM).

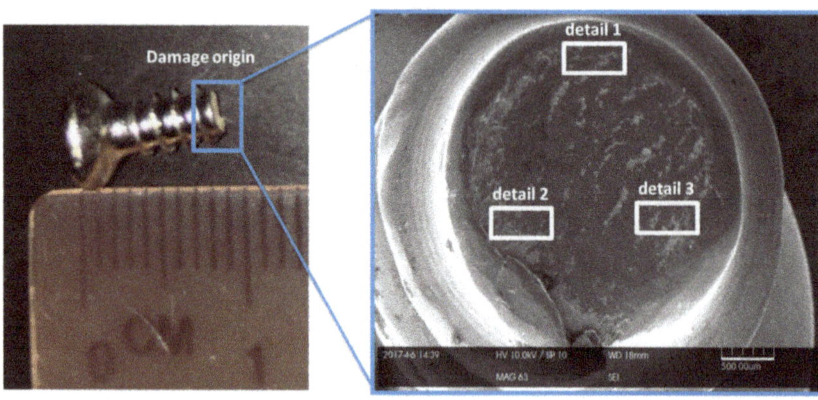

Figure A2. Topographical SEM image of the CS1 at the fracture surface showing the damage origin and the details of interest at the fracture surface.

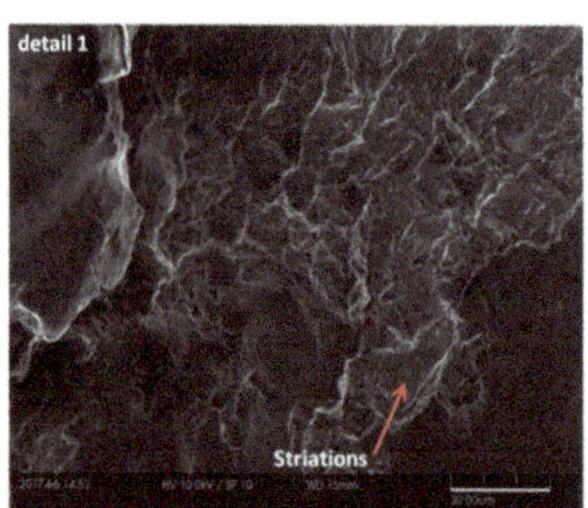

Figure A3. Topographical SEM image of the CS1 showing striations.

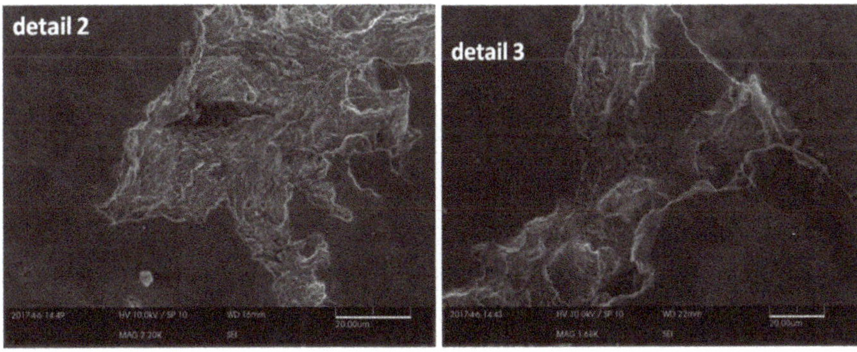

Figure A4. Topographical SEM image of the CS1 showing a microcrack at detail 2 (**left**). Detail 3 (**right**) showing striations with a severely rubbed surface.

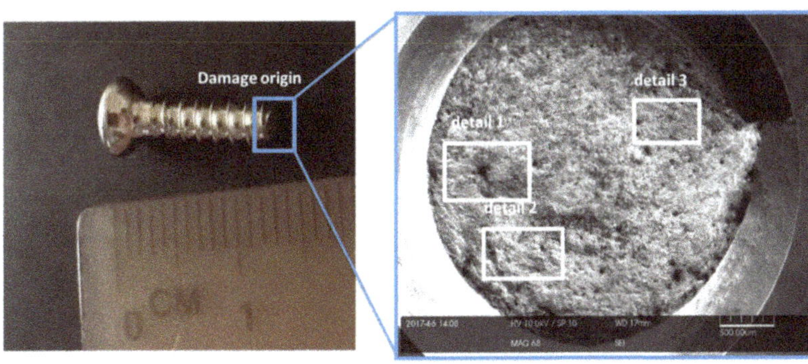

Figure A5. Topographical SEM image of the CS4 showing the details of interest at the fracture surface.

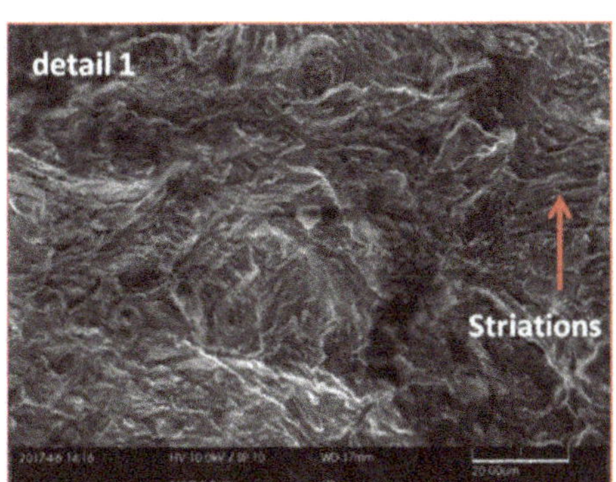

Figure A6. Topographical SEM image of the CS4 showing striations.

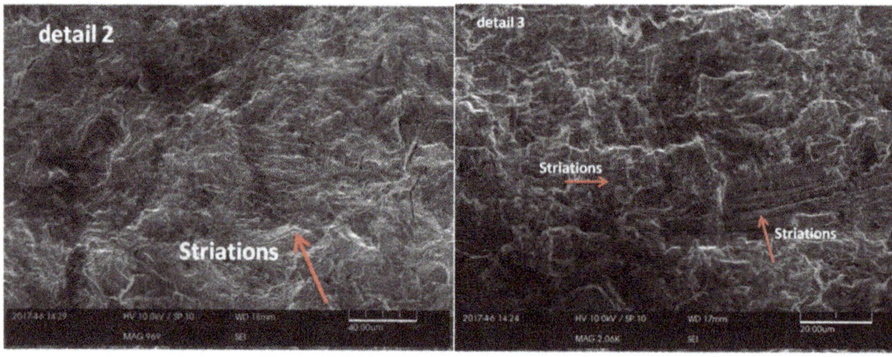

Figure A7. Topographical SEM image of the CS4 showing detail 1 (**left**). Detail 2 (**right**) showing striations.

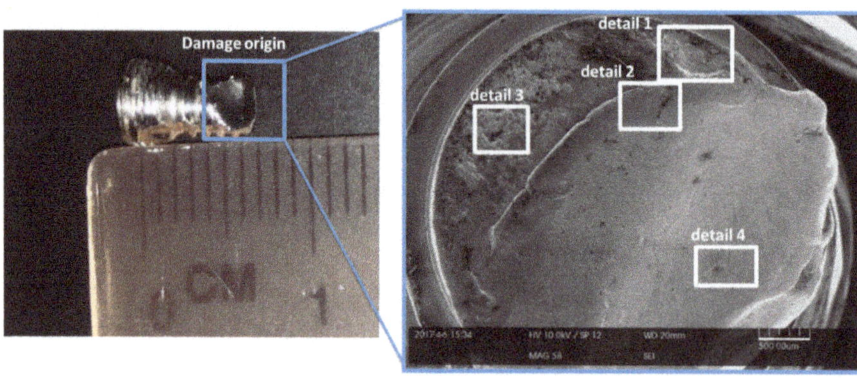

Figure A8. Topographical SEM image of the locking screw (LC) showing the damage origin and the details of interest.

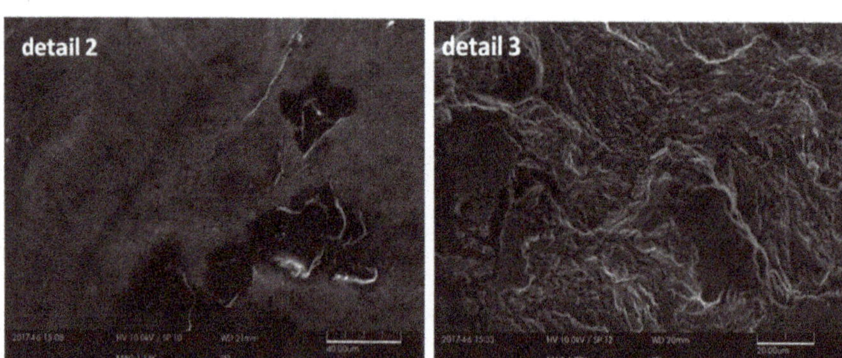

Figure A9. Topographical SEM image of the locking screw (LC) showing highly mechanically rubbed surfaces (**left**) with some striations on detail 3 (**right**).

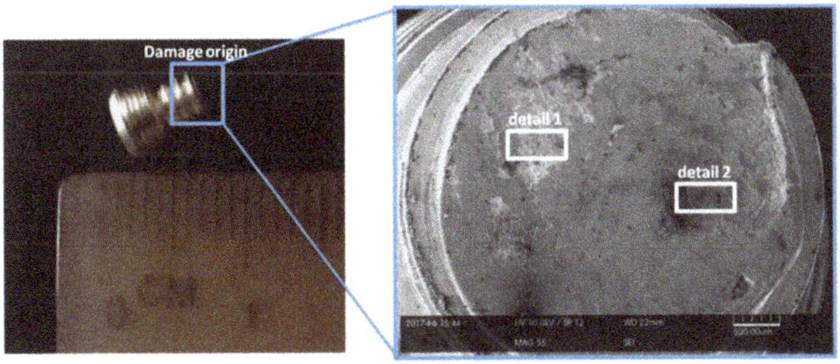

Figure A10. Topographical SEM image of the locking screw (LE) at the fracture surface showing the damage origin and the details of interest.

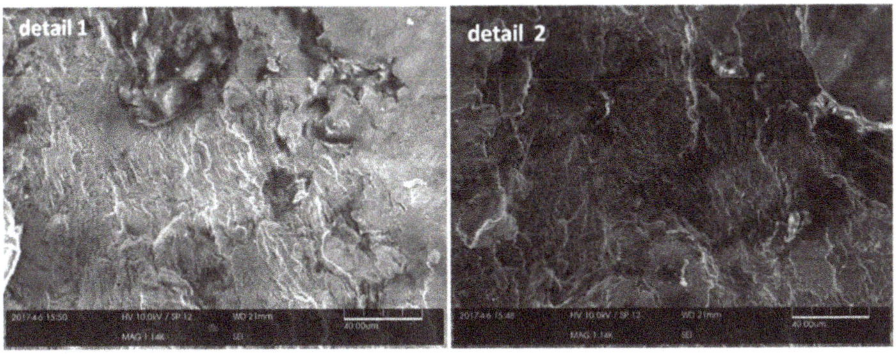

Figure A11. Topographical SEM image of the locking screw (LE) showing detail 1 (**left**). Detail 2 (**right**) showing some striations.

References

1. Ahmad, J.; Pour, A.E.; Raikin, S.M. The modified use of a proximal humeral locking plate for tibiotalocalcaneal arthrodesis. *Foot Ankle Int.* **2007**, *28*, 977–983. [CrossRef] [PubMed]
2. Mendicino, R.W.; Catanzariti, A.R.; Saltrick, K.R.; Dombek, M.F.; Tullis, B.L.; Statler, T.K.; Johnson, B.M. Tibiotalocalcaneal arthrodesis with retrograde intramedullary nailing. *J. Foot Ankle Surg.* **2004**, *43*, 82–86. [CrossRef] [PubMed]
3. Fox, I.M.; Shapero, C.; Kennedy, A. Tibiotalocalcaneal arthrodesis with intramedullary interlocking nail fixation. *Clin. Podiatr. Med. Surg.* **2000**, *17*, 19–31. [PubMed]
4. Shearman, A.D.; Eleftheriou, K.I.; Patel, A.; Pradhan, R.; Rosenfeld, P.F. Use of a proximal humeral locking plate for complex ankle and Hindfoot fusion. *J. Foot Ankle Surg.* **2016**, *55*, 612–618. [CrossRef] [PubMed]
5. Berend, M.E.; Glisson, R.R.; Nunley, J.A. A biomechanical comparison of intramedullary nail and crossed lag screw fixation for tibiotalocalcaneal arthrodesis. *Foot Ankle Int.* **1997**, *18*, 639–643. [CrossRef] [PubMed]
6. Chiodo, C.P.; Acevedo, J.I.; Sammarco, V.J.; Parks, B.G.; Boucher, H.R.; Myerson, M.S.; Schon, L.C. Intramedullary rod fixation compared with blade-plate-and-screw fixation for tibiotalocalcaneal arthrodesis: A biomechanical investigation. *J. Bone Joint Surg. Am.* **2003**, *85*, 2425–2428. [CrossRef] [PubMed]
7. Goswami, T.; Patel, V.; Dalstrom, D.J.; Prayson, M.J. Mechanical evaluation of fourth-generation composite femur hybrid locking plate constructs. *J. Mater. Sci. Mater. Med.* **2011**, *22*, 2139. [CrossRef] [PubMed]
8. Perren, S.M. Technical and biomechanical aspects of screws used in bone surgery. *Int. J. Orthop. Trauma* **1992**, *2*, 31–48.

9. Asnis, S.E.; Kyle, R.F. (Eds.) *Cannulated Screw Fixation: Principles and Operative Techniques*; Springer Science & Business Media: New York, NY, USA, 2012.
10. Patel, V. Biomechanical Evaluation of Locked and Non-Locked Constructs under Axial and Torsion Loading. (Electronic Thesis or Dissertation). 2008. Available online: https://etd.ohiolink.edu/ (accessed on 14 April 2018).
11. Gautier, E.; Sommer, C. Guidelines for the clinical application of the LCP. *Injury* **2003**, *34*, B63–B76. [CrossRef] [PubMed]
12. Hsu, C.C.; Wang, J.L.; Hou, S.M.; Chao, C.K.; Lin, J. Pushout strength of tibial locking screws: Development of finite element models. *J. Chin. Inst. Eng.* **2003**, *26*, 817–823. [CrossRef]
13. Hughes, A.N.; Jordan, B.A. The mechanical properties of surgical bone screws and some aspects of insertion practice. *Injury* **1972**, *4*, 25–38. [CrossRef]
14. Chapman, J.R.; Harrington, R.M.; Lee, K.M.; Anderson, P.A.; Tencer, A.F.; Kowalski, D. Factors affecting the pullout strength of cancellous bone screws. *J. Biomech. Eng.* **1996**, *118*, 391–398. [CrossRef] [PubMed]
15. Shi, Z.; Zhang, C.; Gu, W.; Zeng, B. Ankle arthrodesis by lateral malleolus osteotomy and internal fixation with locking proximal humeral plate. *Zhongguo Xiu Fu Chong Jian Wai Ke Za Zhi (Chin. J. Repar. Reconstr. Surg.)* **2011**, *25*, 781–784.
16. Ina, J.; Vallentyne, M.; Hamandi, F.; Shugart, K.; Boin, M.; Laughlin, R.; Goswami, T. Failure Analysis of PHILOS Plate Construct Used for Pantalar Arthrodesis Paper I—Analysis of the Plate. *Metals* **2018**, *8*, 180. [CrossRef]
17. SYNTHES. 3.5 mm LCP Proximal Humerus Plate—Stainless Steel and Titanium Technique Guide. 2002. Available online: http://synthes.vo.llnwd.net/o16/LLNWMB8/US%20Mobile/Synthes%20North%20America/Product%20Support%20Materials/Technique%20Guides/DSUSTRM10161133_ProxHumPl_STG_150dpi.pdf (accessed on 7 February 2017).
18. Atlas Specialty Metals. Stainless Steel—Grade 316 L—Properties, Fabrication and Applications (UNS S31603). 2004. Available online: http://www.azom.com/article.aspx?ArticleID=2382# (accessed on 7 February 2017).
19. Azevedo, C.R.F. Failure analysis of a commercially pure titanium plate for osteosynthesis. *Eng. Fail. Anal.* **2002**, *10*, 153–164. [CrossRef]
20. Kanchanomai, C.; Phiphobmongkol, V.; Muanjan, P. Fatigue failure of an orthopedic implant–A locking compression plate. *Eng. Fail. Anal.* **2008**, *15*, 521–530. [CrossRef]
21. Amalraju, D.; Dawood, A.S. Mechanical strength evaluation analysis of stainless steel and titanium locking plate for femur bone fracture. *IRACST Eng. Sci. Technol. Int. J.* **2012**, *2*, 2250–3498.
22. ASTM. *Annual Book of ASTM Standards*; Construction; Soil and Rock (I); ASTM: West Conshohocken, PA, USA, 2005; Volume 13.01.
23. Allen, R.F. Standard specification for wrought 18 chromium-14 nickle-2.5 molybdenum stainless steel bar and wire for surgical implants (F138). In *Annual Book of ASTM Standards, Medical Devices and Services*; ASTM: West Conshohocken, PA, USA, 1998.
24. SYNTHES. Screws, Drill Bits, Taps and Guide Wires. 2002. Available online: http://www.cambridgeorthopaedics.com/easytrauma/classification/commonfiles/Synthes%20screws.pdf (accessed on 7 February 2017).
25. Hamandi, F.; Goswami, T. Macrodamage Accumulation Model for a Human Femur. *Appl. Bionics Biomech.* **2017**, *2017*, 4539178. [CrossRef] [PubMed]
26. Michael, J.M.; Golshani, A.; Gargac, S.; Goswami, T. Biomechanics of the ankle joint and clinical outcomes of total ankle replacement. *J. Mech. Behave. Biomed. Mater.* **2008**, *1*, 276–294. [CrossRef] [PubMed]
27. Perren, S.M. Evolution of the internal fixation of long bone fractures: The scientific basis of biological internal fixation: Choosing a new balance between stability and biology. *Bone Jt. J.* **2002**, *84*, 1093–1110. [CrossRef]
28. Goswami, T.; Gundapaneni, D.; Slocum, M.; Paul, P.; Christof, S. Failure investigation of a tibiotalocalcaneal arthrodesis system. *Eng. Fail. Anal.* **2016**, *59*, 588–604. [CrossRef]
29. Egol, K.A.; Kubiak, E.N.; Fulkerson, E.; Kummer, F.J.; Koval, K.J. Biomechanics of locked plates and screws. *J. Orthop. Trauma* **2004**, *18*, 488–493. [CrossRef] [PubMed]
30. Borgeaud, M.; Cordey, J.; Leyvraz, P.F.; Perren, S.M. Mechanical analysis of the bone to plate interface of the LC-DCP and of the PC-FIX on human femora. *Injury* **2000**, *31*, C29–C36. [CrossRef]
31. Narayan, R.; Burt, V.; Lampman, S.; Marken, K.; Marquard, E.; Riley, B. ASM Handbook. In *Materials for Medical Devices*; ASM International: Materials Park, OH, USA, 2012; Volume 23.

32. Majid, K.; Crowder, T.; Baker, E.; Baker, K.; Koueiter, D.; Shields, E.; Herkowitz, H.N. Analysis of in vivo corrosion of 316 L stainless steel posterior thoracolumbar plate systems: A retrieval study. *Clin. Spine Surg.* **2011**, *24*, 500–505.
33. Sivakumar, M.; Mudali, U.K.; Rajeswari, S. Investigation of failures in stainless steel orthopaedic implant devices: Fatigue failure due to improper fixation of a compression bone plate. *J. Mater. Sci. Lett.* **1994**, *13*, 142–145. [CrossRef]
34. Thapa, N.; Prayson, M.; Goswami, T. A failure study of a locking compression plate implant. *Case Stud. Eng. Fail. Anal.* **2015**, *3*, 68–72. [CrossRef]

metals MDPI

Article

In Vitro Corrosion Assessment of Additively Manufactured Porous NiTi Structures for Bone Fixation Applications

Hamdy Ibrahim, AhmadReza Jahadakbar, Amir Dehghan, Narges Shayesteh Moghaddam, Amirhesam Amerinatanzi and Mohammad Elahinia *

Dynamic and Smart Systems Laboratory, Mechanical Industrial and Manufacturing Engineering Department, The University of Toledo, Toledo, OH 43606, USA; hamdy.m.elsayed@gmail.com (H.I.); ar.jahadakbar@gmail.com (A.J.); amirdehghan.g@gmail.com (A.D.); shayesteh.narges@gmail.com (N.S.M.); amirameri83@gmail.com (A.A.)

* Correspondence: mohammad.elahinia@utoledo.edu; Tel.: +1-419-530-8224

Received: 14 January 2018; Accepted: 6 March 2018; Published: 8 March 2018

Abstract: NiTi alloys possess distinct functional properties (i.e., shape memory effect and superelasticity) and biocompatibility, making them appealing for bone fixation applications. Additive manufacturing offers an alternative method for fabricating NiTi parts, which are known to be very difficult to machine using conventional manufacturing methods. However, poor surface quality, and the presence of impurities and defects, are some of the major concerns associated with NiTi structures manufactured using additive manufacturing. The aim of this study is to assess the in vitro corrosion properties of additively manufactured NiTi structures. NiTi samples (bulk and porous) were produced using selective laser melting (SLM), and their electrochemical corrosion characteristics and Ni ion release levels were measured and compared with conventionally fabricated NiTi parts. The additively manufactured NiTi structures were found to have electrochemical corrosion characteristics similar to those found for the conventionally fabricated NiTi alloy samples. The highest Ni ion release level was found in the case of 50% porous structures, which can be attributed to their significantly higher exposed surface area. However, the Ni ion release levels reported in this work for all the fabricated structures remain within the range of most of values for conventionally fabricated NiTi alloys reported in the literature. The results of this study suggest that the proposed SLM fabrication process does not result in a significant deterioration in the corrosion resistance of NiTi parts, making them suitable for bone fixation applications.

Keywords: Nitinol alloys; additive manufacturing; selective laser melting; in vitro corrosion; porous structures; bone implants

1. Introduction

Recently, Nickel-titanium (NiTi) alloys, also known as Nitinol, have become attractive for various applications due to their unique characteristics, such as their shape memory behavior and superelasticity, which allows them to recover from large strains as high as 8% [1–4]. In addition, NiTi alloys are more biocompatible when implanted inside the human body, and have a more desirable strength and modulus of elasticity compared to the metallic alloys conventionally used (e.g., Ti-6Al-4V) [5–8]. The advantageous properties of NiTi alloys make them appealing for different biomedical load-bearing applications, such as staples, clamps, orthodontic wires and bone implants. The high biocompatibility of NiTi alloys is due to their high corrosion resistance, which can be attributed to the fast formation of protective layers on the surface, especially when exposed to Cl-containing environments (e.g., physiological environment). These protective layers mainly consist of TiO, TiO_2 and Ti_2O_3, which are considerably more stable than the NiTi alloy [8–11]. It has been reported that the composition of the protective layers on the surface of NiTi parts depends mainly on the surface characteristics and the composition of the testing solution [12–14]. For instance, the effects of intraoral aging on the surface properties of esthetic and conventional NiTi was studied by Rongo et al. [15]. They found that, even for coated NiTi wires, the surface properties—including the surface roughness—could be altered during and after their clinical use. In addition, the immersion of NiTi alloys in Hank's solution mostly forms TiO_2, $Ca_3(PO_4)_2$ and small traces of $Ni(OH)_2$ as the products of the corrosion process [11,16].

Despite the protection from corrosion provided by the formed layers on the surface of NiTi parts, it has been reported in several studies that NiTi parts could release nickel (Ni) ions when exposed to a physiological environment for long implantation durations [17]. Ni ions are known to be harmful to the surrounding living tissues and to cause allergic reaction and genotoxicity in the human body [11,18,19]. Ryhanen et al. [20] performed a comparative in vitro cell culture study to measure Ni ions released from polished NiTi and electro-polished 316L stainless steel in a cell culture media. In both cases, NiTi samples showed higher Ni ion levels for the first day, with a rapid decrease over the next eight consecutive days until the Ni ion release levels of the 316L stainless steel samples were reached. They found that, despite the higher Ni ion release in non-passivated NiTi samples, the ion concentration value was not high enough to kill the adjacent cells due to toxicity. In a similar set of studies conducted by Wever et al. [16], the Ni ion release of passivated NiTi samples was investigated and compared with 316L stainless steel samples. They found the same results as Ryhanen et al. [20] and reported a Ni ion release rate of $14.5 \times 10^{-7} \mu g \cdot (cm^2 \cdot s)^{-1}$. Likewise, El Medawar et al. [21] investigated the Ni ion release of NiTi samples exposed to different sets of cells and reported that the amount of released Ni ions were 0.098 ± 0.006 ppm and 0.104 ± 0.007 ppm after 3 and 6 days, respectively. The control medium had a constant Ni ion release of 0.01 ppm in the investigation period. In another study, Bernard et al. [22] reported a maximum of 268 ± 11 ppb of Ni ion release in NiTi samples.

NiTi alloys are difficult to process using conventional methods and procedures (e.g., machining). This can be attributed to many reasons, such as the stress-induced martensitic transformation, work hardening, spring-back effects, and burr formation of NiTi alloys [5]. These are the main reasons that the vast majority of practical NiTi devices developed so far have simple and limited geometries, such as wires, tubes, sheets and thin rods. In addition, NiTi is very sensitive to impurity pick up, which makes high-temperature processing challenging [5,23]. These impurities result in the formation of Ti-rich brittle phases, which affect the functional and structural properties of the material. Recently, Additive Manufacturing (AM) methods have been developed for the fabrication of metallic parts, including NiTi [24,25]. This near-net-shaping technology has been able to overcome the manufacturing limitations and has enabled the fabrication of complex geometries, such as scaffolds and porous structures [5,24,26]. In AM methods, the final part is fabricated directly from a Computer Aided Design (CAD) file using layer-by-layer fabrication. This powerful method of fabrication has enabled researchers to design and introduce realistic NiTi implants, actuators, and medical devices with complex geometries that benefit from superelastic and shape memory effects. Several studies have introduced patient-specific, stiffness-matched NiTi bone fixation plates that can be fabricated using additive manufacturing [27,28]. The stiffness matching offers significantly better outcomes for patients and has been achieved by imposing specific levels and types of porosity on the NiTi parts. The use of additive manufacturing in the biomedical field is not limited to the fabrication of metallic parts, but it enables the fabrication of complex shapes for a wide range of materials, including polymers and polymer-based composites [29–31]. For example, Martorelli et al. [32] used additive manufacturing to fabricate customized polymer-based composite mandibles that simulate the mechanical behavior of a human mandible.

The use of AM fabrication techniques and the introduction of porosity to the NiTi parts add other factors that may affect the corrosion characteristics of 3D-printed NiTi alloys and hence the release of the Ni ions in the physiological environment. After an extensive review of the literature, it was found that the corrosion behavior of the additively manufactured NiTi parts has not yet been evaluated. This study focuses on the in vitro evaluation of the corrosion behavior of additively manufactured NiTi parts, fabricated using a selective laser melting (SLM) process, represented in their electrochemical corrosion characteristics and Ni ion release levels. The chosen SLM process provides more freedom to change the level of introduced porosity, pore morphology and material composition of the fabricated NiTi parts needed for bone implant applications. In addition, this work is the first to assess the corrosion and ion release behavior of the porous NiTi parts compared to 1bulk 3D-printed and conventionally fabricated NiTi parts.

2. Materials and Methods

2.1. Design and Fabrication of the Samples

Selective Laser Melting (SLM) was used to fabricate the samples. $Ni_{50.8}Ti_{49.2}$ (at %). Ingots from NiTi Devices and Components, Inc. (Fremont, CA, USA), were provided as the raw material for the SLM process. The ingots were atomized to powder using the Electrode Induction-melting Gas Atomization (EIGA) technique (by TLS Technique GmbH (Bitterfeld, Germany)). Finally, the prepared powder in 25–75 μm particle fractions was used as the input material for the Phenix PXM SLM machine (3D Systems, Valencia, CA, USA). The utilized SLM machine was equipped with a 300 W Ytterbium fiber laser with a beam quality of $M^2 < 1.2$. The beam profile of the machine was Gaussian (TEM_{00}), and the beam diameter was about 80 μm. Table 1 summarizes the process parameters used for fabrication of the parts. More details on the fabrication of the samples using the Phenix machine can be found our previous studies [27,28].

Table 1. Process parameters used in Selective Laser Melting (SLM) manufacturing of NiTi parts.

Effective Laser Power (W)	Layer Thickness (μm)	Scanning Velocity (m/s)	Hatch Distance (μm)	Energy Input (J/mm^3)
250	30	1.25	120	55.5

In order to study the corrosion response of the additively manufactured NiTi parts, bulk and porous cubic samples with dimensions of 9 mm × 9 mm × 9 mm were designed and fabricated. The porous samples were designed by imposing porosity at different levels (15–50%), as shown in Figures 1 and 2. Samples for electrochemical corrosion characterization were fabricated with the same dimensions, with an additional hook connected at the top surface of the cubes for better electrical connectivity during the test. The fabricated samples were then separated from the base by using an EDM cutting machine.

In vitro corrosion assessment of the additively manufactured (AM) NiTi parts (bulk and porous) was performed, and it was compared to the conventionally fabricated NiTi alloy (same atomic composition) using immersion (Ni ion release) and electrochemical corrosion tests. A simulated body fluid (SBF) medium, prepared according to Oyane et al. [33], was used for the in vitro tests, see Table 2. The tested samples were polished using a sand blaster, and the samples for electrochemical corrosion characterization were further polished using SiC papers from 180 to 1500 grit. All the tested samples (immersion and electrochemical tests) were then cleaned by water and ultrasonically cleaned by ethanol before each test.

Table 2. Composition of 1000 mL solution of a modified SBF [33].

Reagent	Amount
NaCl	5.403 g
$NaHCO_3$	0.504 g
Na_2CO_3	0.426 g
KCl	0.225 g
$K_2HPO_4 \cdot 3H_2O$	0.23 g
$MgCl_2 \cdot 6H_2O$	0.311 g
0.2 mol L^{-1} NaOH	100 mL
HEPES	17.892 g
$CaCl_2$	0.293 g
Na_2SO_4	0.072 g
1 mol L^{-1} NaOH	15 mL

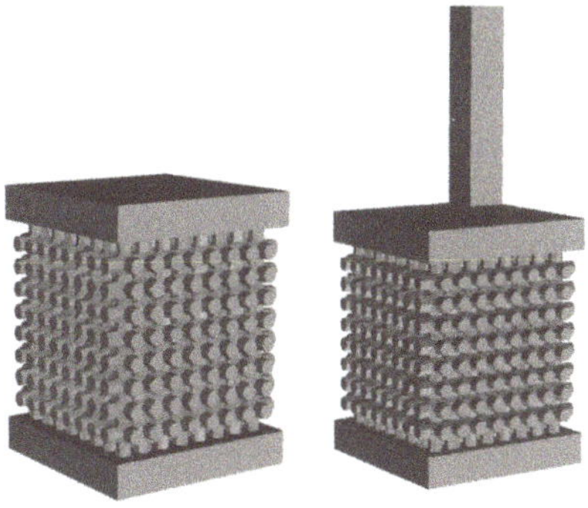

Figure 1. CAD files of the immersion and electrochemical samples.

Figure 2. Fabricated samples for electrochemical tests.

2.2. Electrochemical Corrosion Test

Samples were molded into epoxy resin with a determined area exposed. In the cases of conventionally fabricated and AM samples, the same exposed area of 0.5 cm^2 was chosen for better comparison. In the case of the bulk and porous structures, the whole side surfaces of the part were exposed, and only the top surfaces were covered with epoxy, see Figure 3. In this way, a better comparison between the electrochemical corrosion characteristics of bulk and porous NiTi devices could be determined.

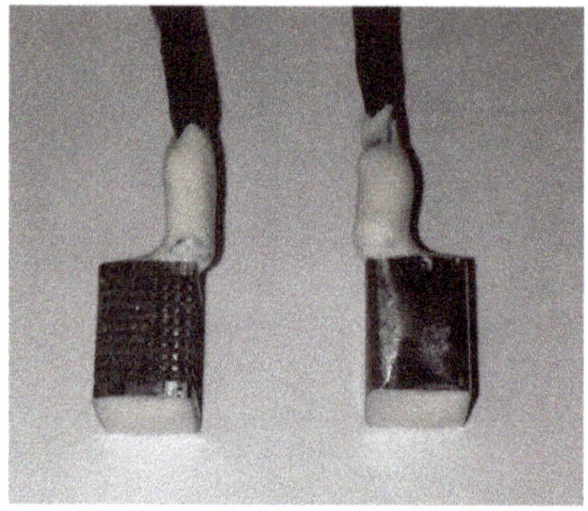

Figure 3. The epoxy molding of the bulk and porous samples and exposed surfaces after the electrochemical corrosion test in SBF.

A Gamry (Gamry, Centennial, CO, USA) Instruments Potentiostat was used to perform the potentiodynamic polarization (PDP) test in the SBF at pH 7.4, 37 ± 1 °C and a constant scan rate of 1 mV/s from −250 to +1000 mV (versus ref. voltage) with respect to the OCP value. A standard three-electrode system with silver chloride (Ag/AgCl) as the reference electrode, graphite rod as the counter electrode, and the sample as the working electrode, see Figure 4, was used. The potential values obtained from the silver chloride reference electrode were properly converted to the standard calomel electrode (SCE) potentials for better comparison with the literature. Open circuit potential (OCP) of the samples was measured for durations of 60 min to attain a stable state prior the experiment. Graphical Tafel analysis was used to obtain the corrosion characteristics such as current densities and corrosion potentials.

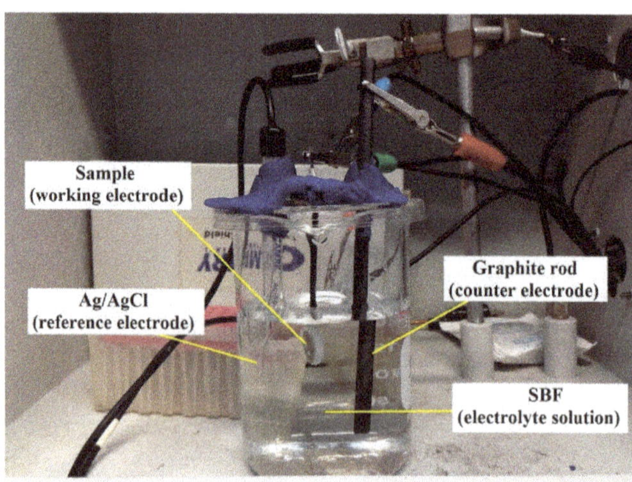

Figure 4. The electrochemical cell unit used for determining the electrochemical characteristics of the prepared samples.

2.3. Surface Morphology

The surface morphology of the NiTi parts after the in vitro potentiodynamic polarization test was investigated using scanning electron microscopy (SEM). A Hitachi (Tokyo, Japan) S-4800 scanning electron microscope equipped with energy dispersive spectroscopy (EDS) analysis was used for this purpose. EDS analysis was used to conduct an elemental analysis on the surface's corrosion products.

2.4. X-ray Diffraction (XRD)

The composition of the surface of the samples before and after the electrochemical corrosion test was investigated using a Rigaku Ultima III X-ray Diffractometer (Rigaku, Tokyo, Japan) with Small Angle X-ray Scattering (SAXS), provided by Rigaku, Tokyo, Japan. XRD with parallel beam (PB) mode was used, and the analysis was acquired for 2θ from 20° to 90° with a step size of 0.05 at 35 mA and 40 kV using Cu Kα radiation.

2.5. Ni Ion Release (Immersion Test)

The immersion test was conducted by submerging the sand-blast polished and ethanol-degreased samples in polypropylene tubes filled with 20 mL SBF solution at pH 7.4 and 37 °C. The temperature was maintained at 37 °C by placing the tubes in a rotary incubator throughout the period of the test (3 days). The continuous rotation of the tubes (at 20 rpm) during the immersion test was used to accelerate the corrosion process and shorten the immersion test duration. In this study, the pH of all the tested samples was recorded daily and the amount of Ni ion released in ppb after 3 days of exposure was measured using Inductively Coupled Plasma Atomic Emission Spectroscopy (ICP-AES). For each tested sample, 1 mL of the SBF medium after 3 days was diluted with deionized water at a ratio of 1:10 in order to produce an adequate sample size and ion concentrations for spectrometry. Calculations were performed according to these dilutions to determine the actual Ni ion release after immersion. The samples were analyzed using the X-series 2 from Thermo Scientific. For quantitative analysis, certified ICP-MS standards from Inorganic Ventures were prepared at 2% HNO_3. Correlation coefficients for calibration curves were above 0.999. The ionic concentrations (mmol/L) of the SBF used in the present study were determined by Hansen et al. [34] and were as follows: 142.0 Na^+, 5.0 K^+, 1.5 Mg^{2+}, 2.5 Ca^{2+}, 147.8 Cl^-, 4.2 HCO_3^-, 1.0 HPO_4^{2-} and 0.5 SO_4^{2-}.

3. Results and Discussion

3.1. Electrochemical Corrosion Test

In vitro corrosion behavior of NiTi plays a significant role in determining the functionality of biomedical devices in the human body. Figure 5 shows the Tafel curves of the conventionally fabricated NiTi alloy and the additively manufactured NiTi alloy. The additively manufactured NiTi alloy shows corrosion characteristics (20 ± 2.1 μA/cm^2 current density and −0.394 ± 0.01 V corrosion potential) comparable to those for the biocompatible conventionally fabricated NiTi alloy (12 ± 3.8 μA/cm^2 current density and −0.4155 ± 0.008 V corrosion potential). The similar corrosion characteristics of the additively manufactured and conventionally fabricated NiTi samples imply homogenization in the microstructure and the absence of apparent defects for the AM NiTi parts that could deteriorate the corrosion properties. The Tafel curves also show a slightly higher corrosion current density for the AM NiTi alloy compared with the conventionally fabricated one. However, such a small difference is within the acceptable range and does not have a significant effect on the corrosion behavior of the additively manufactured NiTi samples. This indicates that the additive manufacturing process does not result in a significant change in the electrochemical corrosion characteristics of NiTi alloys.

Figure 6 shows the Tafel curves of two structures with the same volume produced by additive manufacturing: bulk structure and porous structure (35% porosity level). As expected, the higher surface area and the presence of many edges in the case of the porous structures resulted in a higher corrosion current. For instance, the corrosion current increased from about 200 nA for a bulk structure to be 950 nA for a 35% porous structure. This increase in the corrosion currents is expected to result in more corrosion byproducts and a higher amount of Ni ion released in vivo.

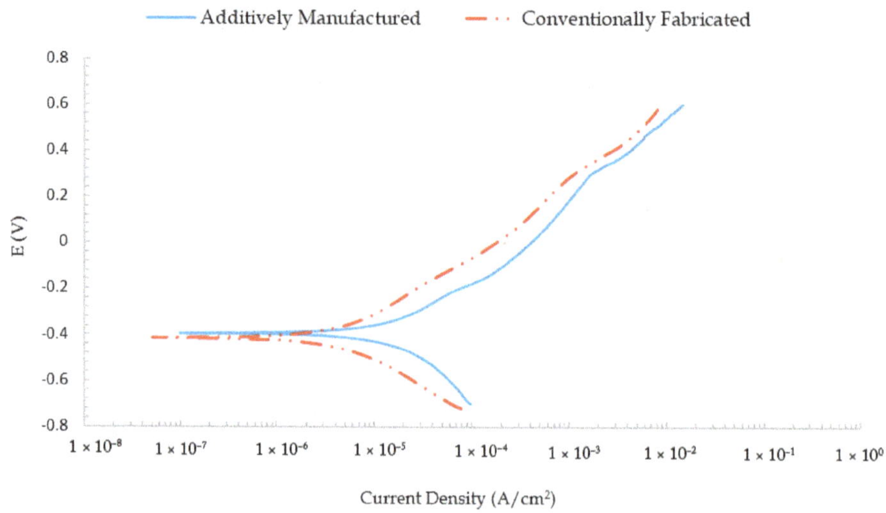

Figure 5. The potentiodynamic polarization curves of the conventionally fabricated and additively manufactured NiTi alloy samples.

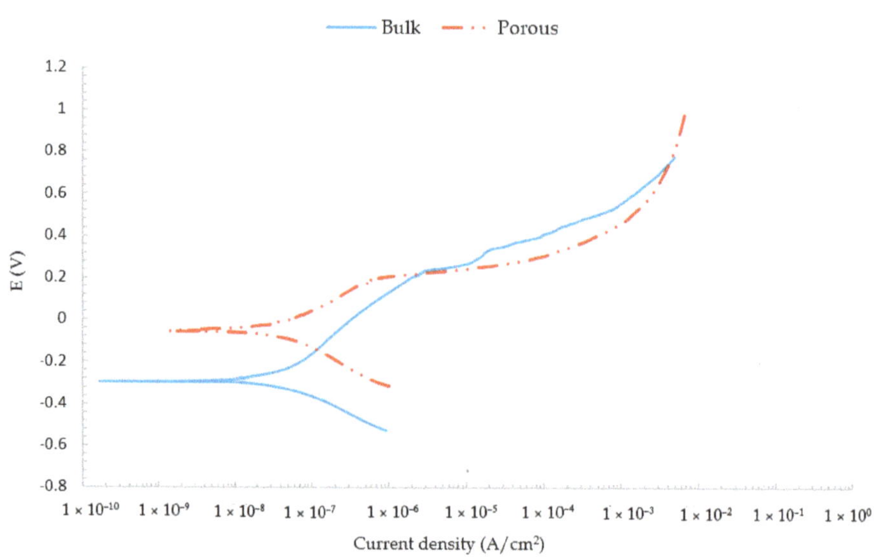

Figure 6. The potentiodynamic polarization curves of the AM NiTi alloys with two different structures; bulk and 30% porous structure.

3.2. *Surface Morphology*

SEM investigation and EDS analysis were used to study the morphology and elemental composition of corroded surfaces for the AM NiTi parts after the in vitro potentiodynamic polarization test, as seen in Figure 7. The corroded surfaces showed several pits with corrosion product agglomerations on the surface. This assures the occurrence of pitting corrosion for the tested samples. The pits observed on the surface of the AM NiTi parts were circular, with a diameter varying between 50 and 250 μm. The EDS elemental analysis showed that the agglomerated corrosion products on the surface of the samples (point 1), concentrated inside and around the pits, consisted mainly of sodium, phosphorous and sulfur-based compounds alongside oxygen, Ti and Ni. However, the elemental analysis away from the pits (point 2) showed no traces of sodium, phosphorus, or sulfur. The presence of oxygen on the surface implies the existence of oxide layers after the corrosion test.

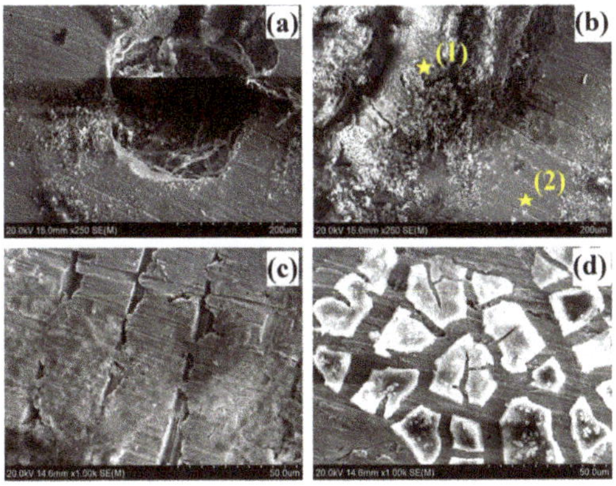

Point	Element Wt.%								
	O	Na	P	S	Cl	K	Ca	Ti	Ni
1	30.4	4.96	1.65	10.39	7.85	1.53	1.53	32.0	8.69
2	10.25	0.0	0.0	0.0	0.0	0.0	0.0	41.56	48.19

Figure 7. SEM investigation and EDS point spectrum analysis of the corroded surface after the potentiodynamic polarization test in SBF at 7.4 pH and 37 °C, (**a**,**b**) show the pits with corrosion product agglomerations at 250× magnification; (**c**,**d**) show the corroded surface at 1000× magnification.

3.3. X-ray Diffraction (XRD)

XRD tests were conducted to determine the compounds available on the surface of the samples before and after the corrosion process. Figure 8 represents the overlapped graphs of XRD of the NiTi samples before and after corrosion. As can be seen, in the graph of additively manufactured NiTi sample, the peaks of NiTi appeared in locations around 28, 43, 61 and 76 degrees, with the second peak being the highest in intensity. However, in the XRD graph of the corroded sample, the NiTi peaks were smaller in intensity, while new phases of HA and TiO_2 emerged. These peaks belong to the corrosion products deposited on the surface of the sample after exposure to the corrosive environment. Moreover, the results of XRD were in a complete agreement with the EDS elemental analysis. Such biocompatible products, in addition to their role in bone growth stimulation, may also act as corrosion barriers that decelerate corrosion.

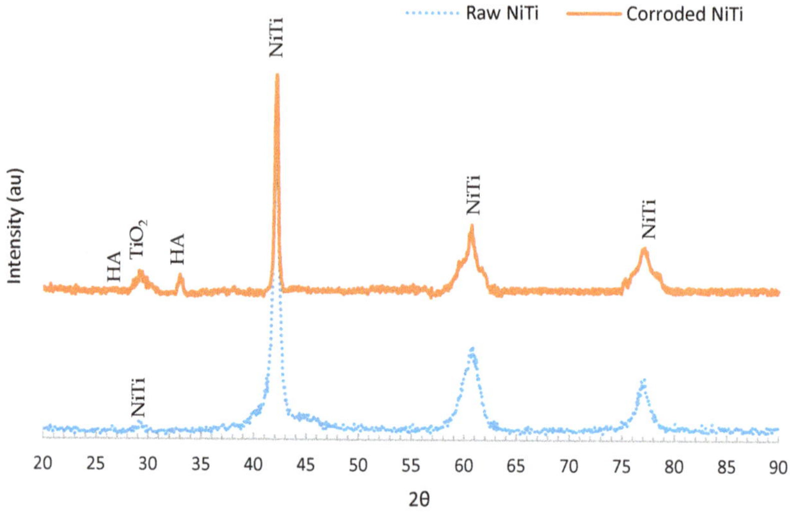

Figure 8. XRD graphs of additively manufactured (raw) and corroded NiTi parts.

3.4. Ni Ion Release (Immersion Test)

The results of the Ni ion release in particles per billion (ppb) after the immersion test can be seen in Figure 9. The results confirm that the amount of the Ni ion released of the AM NiTi (dense structure) was similar to that for the conventionally fabricated NiTi. In addition, the AM porous structures (15, 25, 35 and 50% porosity) showed higher amounts of Ni ion release, which confirms the electrochemical corrosion results. For example, introducing 25% porosity increased the surface area from 594 mm^2 for bulk samples to 1997.5 mm^2 for the porous sample. Such increase in the surface area exposed to the SBF solution led to an increase in the amount of Ni ion release from 64 ppb for bulk samples to 176.35 ppb for the 25% porous structure. However, the maximum Ni ion release of 196.8 ppb in the case of the 50% porous NiTi structure was still within the range of most of the values for conventionally fabricated NiTi alloys reported in the literature [22,35,36]. For instance, Bernard et al. [22] reported a maximum of 268 $\pm$ 11 ppb of Ni ion release for conventionally fabricated NiTi samples. This indicates that both the bulk and porous NiTi structures produced using additive manufacturing had Ni ion release levels similar to those for conventionally fabricated NiTi alloys confirmed to be biocompatible. The surface area calculations and the Ni ion release results are listed in Table 3.

Table 3. The surface area and Ni ion release measurements for the conventionally fabricated and additively manufactured NiTi samples (bulk and porous).

Sample Type	Surface Area (mm^2)	Ni Ions (ppb) (Diluted)	Ni Ions (ppb)
Conventionally-fabricated	590	5.945	59.45
AM (dense)	594	6.4	64
AM (15% porosity)	1620.3	12.79	127.94
AM (25% porosity)	1997.5	17.64	176.35
AM (35% porosity)	2106.6	19.43	194.3
AM (50% porosity)	2116.3	19.68	196.8

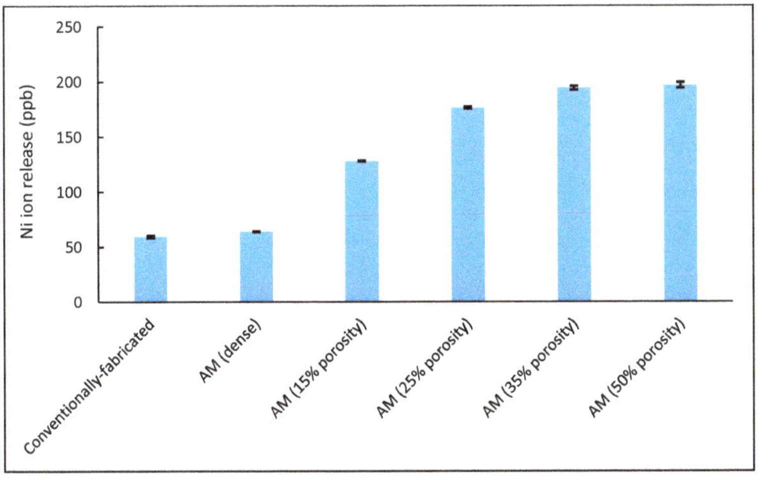

Figure 9. The Ni ion release of two different structures (dense and porous) of additively manufactured NiTi alloys in comparison to the conventionally fabricated NiTi alloy.

4. Conclusions

The additively manufactured NiTi parts were found to have a corrosion behavior (electrochemical corrosion and Ni ion release) similar to that found for the conventionally fabricated NiTi alloy. This indicates that the proposed additive manufacturing process does not result in a deterioration in the corrosion resistance of NiTi parts. In addition, the SEM and XRD analysis of the corroded NiTi samples showed the formation of biocompatible corrosion products on the surface of the sample, which could be beneficial for bone implant applications. Adding porosity to the additively manufactured NiTi parts increased the exposed surface area, which results in a higher electrochemical corrosion current and hence a higher Ni ion release compared to the bulk NiTi samples. However, the Ni ion release levels reported in this work for all the fabricated structures remain within the range of most of values for conventionally fabricated NiTi alloys reported in the literature. It is worth mentioning that there have been several efforts to reduce the Ni ion release of conventionally fabricated NiTi parts in the literature using surface modification and coating techniques. Such techniques could be beneficial to implement for additively manufactured NiTi structures to further increase the corrosion resistance and reduce the Ni ion release rates.

Acknowledgments: The authors wish to acknowledge partial support for this research from Third Frontier (State of Ohio) grant 15-791, titled "Additive Manufacture of Stiffness-Matched Skeletal Fixation Hardware".

Author Contributions: Hamdy Ibrahim and Mohammad Elahinia conceived and designed the experiments; AhmadReza Jahadakbar, Amir Dehghan, Narges Shayesteh Moghaddam, Amirhesam Amerinatanzi and Hamdy Ibrahim performed the experiments; Hamdy Ibrahim and Amir Dehghan analyzed the data; Mohammad Elahinia contributed reagents/materials/analysis tools; Hamdy Ibrahim, Amir Dehghan, AhmadReza Jahadakbar and Mohammad Elahinia wrote the paper.

Conflicts of Interest: The authors declare no conflict of interest.

References

1. Esenwein, S.; Bogdanski, D.; Köller, M.; Krone, L.; Epple, M.; Muhr, G. Clinical applications of shape memory alloys based on niti as implant materials—Possibilities in trauma and orthopaedic surgery. *SMST Proc.* **2006**. [CrossRef]
2. Duerig, T.; Pelton, A.; Stöckel, D. An overview of nitinol medical applications. *Mater. Sci. Eng. A* **1999**, *273*, 149–160. [CrossRef]

3. Peitsch, T.; Klocke, A.; Kahl-Nieke, B.; Prymak, O.; Epple, M. The release of nickel from orthodontic niti wires is increased by dynamic mechanical loading but not constrained by surface nitridation. *J. Biomed. Mater. Res. Part A* **2007**, *82*, 731–739. [CrossRef] [PubMed]
4. Lima de Miranda, R.; Zamponi, C.; Quandt, E. Micropatterned freestanding superelastic tini films. *Adv. Eng. Mater.* **2013**, *15*, 66–69. [CrossRef]
5. Elahinia, M.H.; Hashemi, M.; Tabesh, M.; Bhaduri, S.B. Manufacturing and processing of niti implants: A review. *Prog. Mater. Sci.* **2012**, *57*, 911–946. [CrossRef]
6. Shabalovskaya, S.A. On the nature of the biocompatibility and on medical applications of niti shape memory and superelastic alloys. *Bio-Med. Mater. Eng.* **1996**, *6*, 267–289.
7. Rahmanian, R.; Moghaddam, N.S.; Haberland, C.; Dean, D.; Miller, M.; Elahinia, M. Load bearing and stiffness tailored niti implants produced by additive manufacturing: A simulation study. In *SPIE Smart Structures and Materials + Nondestructive Evaluation and Health Monitoring: International Society for Optics and Photonics*; SPIE: Bellingham, WA, USA, 2014; pp. 905814–905818.
8. Stergioudi, F.; Vogiatzis, C.; Pavlidou, E.; Skolianos, S.; Michailidis, N. Corrosion resistance of porous niti biomedical alloy in simulated body fluids. *Smart Mater. Struct.* **2016**, *25*, 095024. [CrossRef]
9. Hu, T.; Chu, C.; Xin, Y.; Wu, S.; Yeung, K.W.; Chu, P.K. Corrosion products and mechanism on niti shape memory alloy in physiological environment. *J. Mater. Res.* **2010**, *25*, 350–358. [CrossRef]
10. Hu, T.; Xin, Y.; Wu, S.; Chu, C.; Lu, J.; Guan, L.; Chen, H.; Hung, T.F.; Yeung, K.; Chu, P.K. Corrosion behavior on orthopedic niti alloy with nanocrystalline/amorphous surface. *Mater. Chem. Phys.* **2011**, *126*, 102–107. [CrossRef]
11. Figueira, N.; Silva, T.; Carmezim, M.; Fernandes, J. Corrosion behaviour of niti alloy. *Electrochim. Acta* **2009**, *54*, 921–926. [CrossRef]
12. Tan, L.; Dodd, R.; Crone, W. Corrosion and wear-corrosion behavior of niti modified by plasma source ion implantation. *Biomaterials* **2003**, *24*, 3931–3939. [CrossRef]
13. Xu, J.; Jin, X.; Luo, J.; Zhong, Z. Fabrication and properties of porous niti alloys by microwave sintering for biomedical applications. *Mater. Lett.* **2014**, *124*, 110–112. [CrossRef]
14. Cheng, F.; Shi, P.; Man, H. A preliminary study of TiO_2 deposition on niti by a hydrothermal method. *Surf. Coat. Technol.* **2004**, *187*, 26–32. [CrossRef]
15. Rongo, R.; Ametrano, G.; Gloria, A.; Spagnuolo, G.; Galeotti, A.; Paduano, S.; Valletta, R.; D'Antò, V. Effects of intraoral aging on surface properties of coated nickel-titanium archwires. *Angle Orthod.* **2013**, *84*, 665–672. [CrossRef] [PubMed]
16. Wever, D.; Veldhuizen, A.; De Vries, J.; Busscher, H.; Uges, D.; Van Horn, J. Electrochemical and surface characterization of a nickel-titanium alloy. *Biomaterials* **1998**, *19*, 761–769. [CrossRef]
17. Ma, C.; Andani, M.T.; Qin, H.; Moghaddam, N.S.; Ibrahim, H.; Jahadakbar, A.; Amerinatanzi, A.; Ren, Z.; Zhang, H.; Doll, G.L. Improving surface finish and wear resistance of additive manufactured nickel-titanium by ultrasonic nano-crystal surface modification. *J. Mater. Process. Technol.* **2017**, *249*, 433–440. [CrossRef]
18. Wu, H.; Wang, T.; Liu, X.; Lin, N.; Liu, X.; He, Z.; Wang, Z. Fabrication and corrosion resistance of the ti-rich alloyed layer on the surface of niti alloys. *Int. J. Electrochem. Sci.* **2017**, *12*, 2376–2388. [CrossRef]
19. Bolat, G.; Mareci, D.; Iacoban, S.; Cimpoesu, N.; Munteanu, C. The estimation of corrosion behavior of niti and nitinb alloys using dynamic electrochemical impedance spectroscopy. *J. Spectrosc.* **2012**, *2013*, 714920. [CrossRef]
20. Ryhänen, J.; Niemi, E.; Serlo, W.; Niemelä, E.; Sandvik, P.; Pernu, H.; Salo, T. Biocompatibility of nickel-titanium shape memory metal and its corrosion behavior in human cell cultures. *J. Biomed. Mater. Res. Part A* **1997**, *35*, 451–457. [CrossRef]
21. El Medawar, L.; Rocher, P.; Hornez, J.-C.; Traisnel, M.; Breme, J.; Hildebrand, H. Electrochemical and cytocompatibility assessment of nitinol memory shape alloy for orthodontic use. *Biomol. Eng.* **2002**, *19*, 153–160. [CrossRef]
22. Bernard, S.A.; Balla, V.K.; Davies, N.M.; Bose, S.; Bandyopadhyay, A. Bone cell-materials interactions and ni ion release of anodized equiatomic niti alloy. *Acta Biomater.* **2011**, *7*, 1902–1912. [CrossRef] [PubMed]
23. Andani, M.T.; Moghaddam, N.S.; Haberland, C.; Dean, D.; Miller, M.J.; Elahinia, M. Metals for bone implants. Part 1. Powder metallurgy and implant rendering. *Acta Biomater.* **2014**, *10*, 4058–4070. [CrossRef] [PubMed]

24. Wang, X.; Xu, S.; Zhou, S.; Xu, W.; Leary, M.; Choong, P.; Qian, M.; Brandt, M.; Xie, Y.M. Topological design and additive manufacturing of porous metals for bone scaffolds and orthopaedic implants: A review. *Biomaterials* **2016**, *83*, 127–141. [CrossRef] [PubMed]
25. Ibrahim, H.; Esfahani, S.N.; Poorganji, B.; Dean, D.; Elahinia, M. Resorbable bone fixation alloys, forming, and post-fabrication treatments. *Mater. Sci. Eng. C* **2017**, *70*, 870–888. [CrossRef] [PubMed]
26. Amerinatanzi, A.; Moghaddam, N.S.; Ibrahim, H.; Elahinia, M. The effect of porosity type on the mechanical performance of porous niti bone implants. In Proceedings of the ASME 2016 Conference on Smart Materials, Adaptive Structures and Intelligent Systems, San Antonio, TX, USA, 10–12 September 2016; American Society of Mechanical Engineers: New York, NY, USA, 2016; p. V002T006A018.
27. Jahadakbar, A.; Shayesteh Moghaddam, N.; Amerinatanzi, A.; Dean, D.; Karaca, H.E.; Elahinia, M. Finite element simulation and additive manufacturing of stiffness-matched niti fixation hardware for mandibular reconstruction surgery. *Bioengineering* **2016**, *3*, 36. [CrossRef] [PubMed]
28. Moghaddam, N.S.; Skoracki, R.; Miller, M.; Elahinia, M.; Dean, D. Three dimensional printing of stiffness-tuned, nitinol skeletal fixation hardware with an example of mandibular segmental defect repair. *Procedia CIRP* **2016**, *49*, 45–50. [CrossRef]
29. Youssef, A.; Hollister, S.J.; Dalton, P.D. Additive manufacturing of polymer melts for implantable medical devices and scaffolds. *Biofabrication* **2017**, *9*, 012002. [CrossRef] [PubMed]
30. Ibrahim, H.; Mehanny, S.; Darwish, L.; Farag, M. A comparative study on the mechanical and biodegradation characteristics of starch-based composites reinforced with different lignocellulosic fibers. *J. Polym. Environ.* **2017**, 1–14. [CrossRef]
31. Ibrahim, H.; Klarner, A.D.; Poorganji, B.; Dean, D.; Luo, A.A.; Elahinia, M. Microstructural, mechanical and corrosion characteristics of heat-treated Mg-1.2 Zn-0.5 ca (wt %) alloy for use as resorbable bone fixation material. *J. Mech. Behav. Biomed. Mater.* **2017**, *69*, 203–212. [CrossRef] [PubMed]
32. Martorelli, M.; Maietta, S.; Gloria, A.; De Santis, R.; Pei, E.; Lanzotti, A. Design and analysis of 3d customized models of a human mandible. *Procedia CIRP* **2016**, *49*, 199–202. [CrossRef]
33. Oyane, A.; Kim, H.M.; Furuya, T.; Kokubo, T.; Miyazaki, T.; Nakamura, T. Preparation and assessment of revised simulated body fluids. *J. Biomed. Mater. Res. Part A* **2003**, *65*, 188–195. [CrossRef] [PubMed]
34. Hansen, A.W.; Führ, L.T.; Antonini, L.M.; Villarinho, D.J.; Marino, C.E.B.; Malfatti, C.D.F. The electrochemical behavior of the niti alloy in different simulated body fluids. *Mater. Res.* **2015**, *18*, 184–190. [CrossRef]
35. Ramazanzadeh, B.A.; Ahrari, F.; Sabzevari, B.; Habibi, S. Nickel ion release from three types of nickel-titanium-based orthodontic archwires in the as-received state and after oral simulation. *J. Dent. Res. Dent. Clin. Dent. Prospect.* **2014**, *8*, 71.
36. Azizi, A.; Jamilian, A.; Nucci, F.; Kamali, Z.; Hosseinikhoo, N.; Perillo, L. Release of metal ions from round and rectangular niti wires. *Prog. Orthod.* **2016**, *17*, 10. [CrossRef] [PubMed]

metals

MDPI

Article

Design, Modeling, Additive Manufacturing, and Polishing of Stiffness-Modulated Porous Nitinol Bone Fixation Plates Followed by Thermomechanical and Composition Analysis

Ahmadreza Jahadakbar [1], Mohammadreza Nematollahi [1], Keyvan Safaei [1], Parisa Bayati [1], Govind Giri [1], Hediyeh Dabbaghi [1], David Dean [2] and Mohammad Elahinia [1,*]

1 Department of Mechanical, Industrial, and Manufacturing Engineering, The University of Toledo, Toledo, OH 43606, USA; Ahmadreza.Jahadakbar@rockets.utoledo.edu (A.J.); mnemato@rockets.utoledo.edu (M.N.); Keyvan.SafaeiBaghbaderani@rockets.utoledo.edu (K.S.); Parisa.BayatiMalayeri@rockets.utoledo.edu (P.B.); Govind.Giri@rockets.utoledo.edu (G.G.); hdabbag@rockets.utoledo.edu (H.D.)

2 Department of Plastic and Reconstruction Surgery, The Ohio State University, Columbus, OH 43210, USA; David.Dean@osumc.edu

* Correspondence: mohammad.elahinia@utoledo.edu; Tel.: +1-419-530-8224

Received: 30 December 2019; Accepted: 16 January 2020; Published: 19 January 2020

Abstract: The use of titanium bone fixation plates is considered the standard of care for skeletal reconstructive surgery. Highly stiff titanium bone fixation plates provide immobilization immediately after the surgery. However, after the bone healing stage, they may cause stress shielding and lead to bone resorption and failure of the surgery. Stiffness-modulated or stiffness-matched Nitinol bone fixation plates that are fabricated via additive manufacturing (AM) have been recently introduced by our group as a long-lasting solution for minimizing the stress shielding and the follow-on bone resorption. Up to this point, we have modeled the performance of Nitinol bone fixation plates in mandibular reconstruction surgery and investigated the possibility of fabricating these implants. In this study, for the first time the realistic design of stiffness-modulated Nitinol bone fixation plates is presented. Plates with different levels of stiffness were fabricated, mechanically tested, and used for verifying the design approach. Followed by the design verification, to achieve superelastic bone fixation plates we proposed the use of Ni-rich Nitinol powder for the AM process and updated the models based on that. Superelastic Nitinol bone fixation plates with the extreme level of porosity were fabricated, and a chemical polishing procedure used to remove the un-melted powder was developed using SEM analysis. Thermomechanical evaluation of the polished bone fixation plates verified the desired superelasticity based on finite element (FE) simulations, and the chemical analysis showed good agreement with the ASTM standard.

Keywords: Nitinol; additive manufacturing; patient specific implant; bone fixation plate; porous structure; stiffness-modulation

1. Introduction

Oral cancer, infection, tumors, and trauma may lead to structural defects in the mandible either aesthetically or functionally, which needs to be treated through reconstructive surgery. In this surgery, the defective bone is removed and replaced with a bone graft using bone fixation plates and screws. Graft mispositioning causes serious problems and so the reconstruction of segmental bony defects needs to be accurate to provide facial contour with proper height and width of the bone and restore jaw continuity [1]. To this end, fixation plates, most commonly made of Ti-6AL-4V (Ti64), or generally known as titanium), are used to provide stabilization and immobilization of the bone graft. Although

Ti64 bone fixation plates provide a high level of immobilization immediately after the surgery, they may lead to stress shielding (bone resorption) or stress concentrations (device failure). Due to the stiffness-mismatch between the Ti64 bone plates and the bone, stress shielding occurs on the adjacent bone tissue, based on the bone remodeling theory, which leads to bone resorption. Followed by continuous bone resorption on the adjacent bone, the stress levels on the bone plates increase and this may lead to the failure of the plate, bone, and the surgery.

As a solution, our group has proposed and introduced a new generation of bone fixation plates that are stiffness matched to the bone tissue and are fabricated via additive manufacturing [2–7]. Stiffness modulation and the matching is achieved via replacing the Ti64 with Nitinol, as well as incorporating engineered porosity to the bone fixation plates. Nitinol (NiTi), is a biocompatible, low stiffness, shape memory alloy, which benefits from interesting features such as superelasticity and shape memory. Nitinol has been already used in many industrial applications as well as biomedical applications [8–14]. In addition to the inherent low stiffness, NiTi superelastic behavior can be made to be very similar to bone tissue. This makes it a great candidate for bone fixation and other skeletal reconstruction applications. Although NiTi has a relatively lower level of stiffness, it still provides a higher level of stiffness in comparison to the bone tissue. We have shown that by introducing an engineered level and type of porosity to a bone fixation plate, one is able to further reduce the stiffness of NiTi bone fixation plates and reach the level of bone tissue [15,16]. We have also shown that the additive manufacturing method, in the form of selective laser melting (SLM), can be used for the fabrication of porous NiTi bone fixation plates [17,18]. Although we have shown successful results in different sections of this novel approach, in this article with a focus on the fabrication approach (i.e., additive manufacturing), we characterized the fabricated parts and updated the design methodology based on that.

Advances in AM techniques and process development, we have been encouraged by many researchers to study the specific fabrication of biomedical components, which were hard or impossible to fabricate by conventional methods (e.g., machining, forging, etc.) in some cases. The use of AM allows many opportunities not open to conventional methods, such as patient-specific implants with tailorable properties [19–22], complex geometries [16], and better accuracies [23]. Rana et al. fabricated a titanium patient-specific implant for reconstruction of the unilateral orbital fracture using selective laser melting and compared it with a pre-bent titanium mesh, which was manually bent to the desired shape. The results of 34 cases showed a higher degree of accuracy of reconstruction in the case of SLM fabricated implants [23]. A custom made SLM titanium implant used in [24] to treat post-traumatic zygomatic deficiency showed no sign of infection after one year with good integration, suggesting SLM implants are an effective approach for alloplastic craniomaxillofacial bone reconstruction. A clinical study of ten patients who used 3D-printed patient-specific Ti plates showed facilitation in jaw reconstructive surgery as well as higher accuracies [25]. Most of the research in this area has been conducted on Ti-6AL-4V plates and/or implants. In recent years, additive manufacturing of NiTi alloys are getting more attention for the fabrication of complex shapes and geometries [26–30]. This is mostly due to the freedom of fabrication and the superior properties of NiTi, as mentioned earlier. However, most of the research is fundamental, aimed at finding optimal fabrication process parameters and their effects on the part's properties [31], lattice structures [18], corrosion behavior [3], modeling [32], and biocompatibility [33]. It should also be noted that all the SLM fabricated porous structures in the literature have been evaluated only in compression mode and no study, as far as we know, has been done on a realistic stiffness-matched porous bone fixation plate, which is under tension.

In this paper, standard-shaped 4-hole bone fixation plates with modulated levels of stiffness were modeled and simulated in a finite element (FE) model. To validate the design and modeling procedure, designed bone fixation plates with different levels of porosity were fabricated via selective laser melting using Ni50.1Ti powder and mechanically tested under tension. After validation of the modeling procedure, to achieve superelastic behavior, Ni-rich Ni50.8Ti powder was used for the fabrication of the second generation of the bone fixation plates. Thermomechanical and composition analysis of the superelastic stiffness-matched porous bone fixation plates were discussed and a proper methodology for polishing and removing the un-melted powder is then proposed.

2. Materials and Methods

In order to test the bone fixation plates using a standard tensile setup, a straight (not curved) standard 4-hole bone fixation plate geometry, as shown in Figure 1a, was used for the fabrication of the bone fixation plates in this paper. For the first series of fabrication, bone fixation plates with a 3-mm thickness profile were designed and fabricated. At the second stage of the project, after validating the FE model, to characterize the bone fixation plates with the critical level of porosity and the most complex geometry, bone fixation plates with 1.5 mm thickness were fabricated. In order to impose porosity on the bone fixation plates, a cubic pore cell as shown in Figure 1b, was used. The cube was 1 mm and by changing the thickness of the sides, the level of porosity was modulated. A thin 0.5 mm cover as shown in Figure 1 was also designed to cover the bone fixation plate and provide the required support for screws.

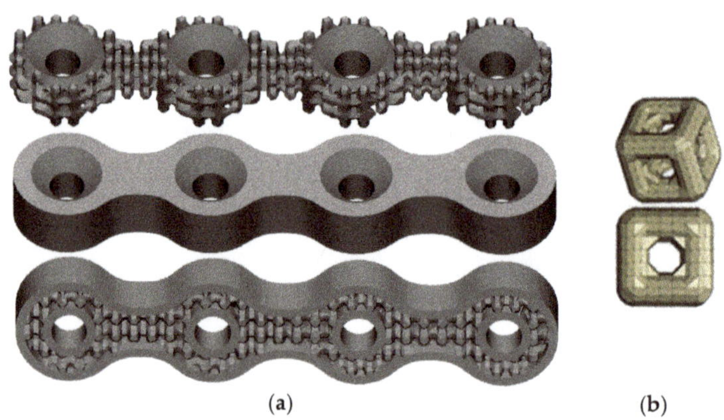

Figure 1. (**a**) Overall geometry of the fabricated bone fixation plates and (**b**) unit pore cell used for creating the porous bone fixation plates.

Two batches of Ni50.1Ti (at. %) and Ni50.8Ti (at. %) ingot were gas atomized using an electrode induction-melting gas atomization (EIGA) by TLS Technik GmbH (Bitterfeld, Germany) to make powder. The powder was sieved to produce particle sizes from 20 to 75 μm. The components of NiTi were produced using an SLM machine (Phenix Systems PXM, [3D Systems], Rock Hill, SC, USA). Our PXM SLM machine has a 300 W Ytterbium fiber laser with a wavelength of 1070 nm and a spot diameter of 80 μm that yields a Gaussian (TEM00) profile. To lessen the impurities in the manufactured samples, the oxygen level in the fabrication chamber was controlled to be less than 800 ppm. There are two series of SLM fabrication process parameters, which are shown in Table 1. The standard dog-bone samples of Ni50.8Ti with the gauge length of 20 mm, gauge the width of 3 mm, and thickness of 2 mm were fabricated for capturing the mechanical properties of as-fabricated SLM parts.

Table 1. The process parameters (PPs) of two series of selective laser melting (SLM) fabrication.

ID	Powder	Laser Power (P; W)	Scanning Speed (V; mm/s)	Hatch Spacing (H; μm)	Layer Thickness (t; μm)
#1	Ni50.1Ti	250	1250	120	30
#2	Ni50.8Ti	250	1250	80	30

A Perkin-Elmer (Waltham, MA, USA) Pyris 1 differential scanning calorimetry (DSC) was used to identify the transformation temperatures (TTs) from −90 to 100 °C with a heating/cooling rate of 10 °C /min in a nitrogen atmosphere by separating small portions of samples ranging from 30 to 45 mg. The standard tensile samples were tested with 25 kN TestResources 910 Series Servohydraulic fatigue test machine (TestResources, Shakopee, MN, USA). A BOSE ElectroForce 3330 machine (TA Instruments, New Castle, DE, USA) was employed to test the bone fixation plates. All samples were loaded with a strain rate of 10^{-4} (s^{-1}) to make sure of isothermal condition. All tests were performed at room temperature (RT). A 2D digital image correlation (DIC) system (correlated solutions, Irmo, SC, USA) by Correlated Solutions, which uses a 5-megapixel camera, a Tonika 100 mm lens, and VIC_2D software was employed to measure the strain. The samples were painted with black and white speckle pattern for DIC measurement. The chemical analyses were performed on the SLM bone fixation plates according to ASTM 2063 standards (Table 2). Two different etching solutions (Table 3) and three different exposing times (2, 4, and 6 min) were utilized to chemically polish the superelastic porous bone fixation plates [34–36]. To evaluate the polishing procedure a FEI Quanta 3D FEG scanning electron microscopy (SEM, Thermo Scientific, Waltham, MA, USA) was used.

Table 2. ASTM 2063 test method description.

Test Method	Description
ASTM E1409-13	Inert gas fusion for oxygen and nitrogen
ASTM R146-83	Vacuum hot extraction for hydrogen
ASRM E1941-16	Combustion infrared detection for carbon
ASTM E1097-12	Direct current plasma emission spectroscopy for all other elements

Table 3. The chemical composition of two etching solutions used for chemical polishing.

Etching Solution	Chemical Composition	Ratio ($HF:HNO_3:H_2O$)
Solution 1	HF (10%), HNO_3 (40%), H_2O (50%)	1:4:5
Solution 2	HF (5%), HNO_3 (50%), H_2O (45%)	1:10:9

3. Modeling

The bone fixation plates that we prepared were 3 mm thick with five different levels of porosity (bulk (0%), 17%, 20%, 24%, 27%, and 30%) were designed based on the methodology explained in the methods section. The STL file of designed bone fixation plates was re-meshed and proper volume meshes (10-node tetrahedral elements (C3D10)) were created and verified in a mesh convergence study. The final volume meshes were exported as INP files and imported to Abaqus (V2019, Dassault Systems; Providence, RI, the USA for finite element modeling). An Abaqus user-defined material subroutine (UMAT) developed based on the micro-plane theory [37–39] was used for capturing the mechanical behavior of the NiTi bone fixation plates. The UMAT is capable of simulating both superelastic and shape memory effects of shape memory alloys under multiaxial loading conditions [15] and requires the thermomechanical properties of the standard coupons (e.g., Young's modulus for austenite and martensite, transformation temperatures, critical stresses, etc.) as the input. More details on the utilized UMAT can be found elsewhere [40].

Thermomechanical properties of Ni50.1Ti from [41] were used for tuning the UMAT mentioned above. Followed by preparing the FE model, the uniaxial tensile test of the bone fixation plates was simulated. The load-displacement response of the bone fixation plates under tensile loading is summarized in Figure 2. As it is seen in Figure 2, adding different levels of porosity to the bone fixation plates alters the stiffness of the plates and can be used for achieving engineered levels of stiffness. As the next step, the simulated bone fixation plates were fabricated and used for validating the FE simulations.

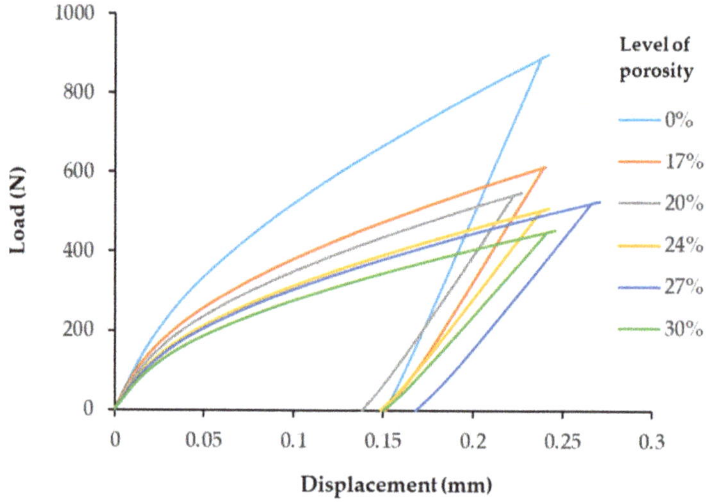

Figure 2. The load–displacement response of simulated plates with different levels of porosity.

4. Mechanical Evaluation

To validate the simulation results, a series of uniaxial tensile mechanical testing on the first series of bone fixation plates with different levels of porosity was performed. All the parts were fabricated horizontally and on a support structure. As shown in Figure 3, the same trend was observed for the load-displacement response of the fabricated bone fixation plates, so that the stiffness of the plate decreased as the level of porosity goes up. For making a comparison between experimental and simulation data, the results of two extreme conditions (bulk bone fixation plate, 30% porosity) were plotted in Figure 4, and as it can be seen, the simulation results are in good agreement with experimental data.

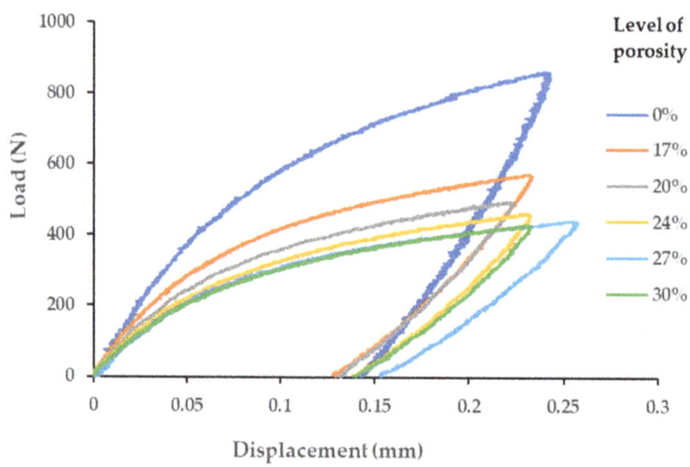

Figure 3. The experimental result of an SLM NiTi fixation plate with different porosity levels under tension.

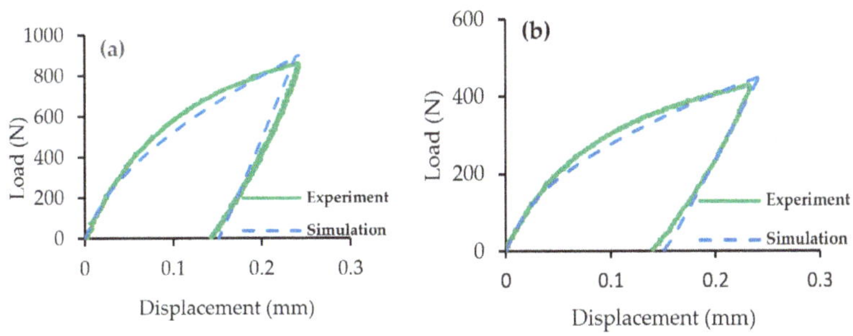

Figure 4. Comparison of experimental and simulation results of (**a**) bulk plate (0% of porosity) and (**b**) the plate with the porosity of 30%.

As was shown in the previous sections, the stiffness of the bone fixation plates, regardless of their overall geometry, can be modulated and tuned to a specific level based on the stiffness of the bone to which it will be attached. However, based on our reading of the clinical literature, it would be useful to be able to create superelastic bone fixation plates in order to recover a larger level of deformation and insure compression between adjacent bone segments that are healing. The superelasticity of bone fixation plates could also act as a shock absorber to prevent failure of the implant in case of short term high impact forces or extreme, but unusual, loading conditions (i.e., failsafe). In addition, it is sometimes useful to apply constant pressure where the bone fixation plates are used. Contact pressure (compression) created by the bone fixation plate should improve the healing procedure and reduce the micromotions that would rip apart the healing region. However, as it can be seen in the load-displacement response (Figures 3 and 4), the fabricated bone fixation plates did not exhibit superelasticity and a noticeable level of recoverable strain. Therefore, after validation of the design, modeling, and fabrication procedure, we replaced the Ni50.1Ti powder used for the SLM procedure with a more Ni-rich powder (Ni50.8Ti) to reduce the transformation temperatures and create superelastic bone fixation plates. The next section discussed the details of the transition to superelastic bone fixation plates.

5. Superelastic Stiffness-Modulated Bone Fixation Plates

In order to investigate the superelastic response of bone fixation plates, only bone fixation plates with the most complex geometry (minimum thickness = 1.5 mm) and an extreme level of porosity (i.e., 46%) was considered. To create a realistic geometry for the bone fixation plate, a 4-hole 1.5 mm thick bone fixation plate as shown in Figure 5a was provided by Stryker (Kalamazoo, MI, USA). The plate was then scanned using a micro computed tomography (micro-CT), and its CAD model was created (Figure 5b). Using the same methodology explained earlier, the porosity (0.6 mm pore cells) and a covering thin-wall structure (0.2 mm) added to the CAD model (Figure 5c,d). After creating the final CAD model for the porous plate, by following the previous procedure, an FE model for simulating the tensile behavior as well as an STL file for the SLM fabrications was created. In order to accurately characterize the tensile behavior of the Ni50.8Ti bone fixation plates, standard tensile samples were also designed and prepared for fabrication in addition to the bone fixation plates. Figure 6 shows the fabricated bone fixation plates immediately after fabrication and before the support removal procedure. Localized and overall thermomechanical behavior of the standard tensile coupon as well as the bone fixation plates, is discussed in the next section.

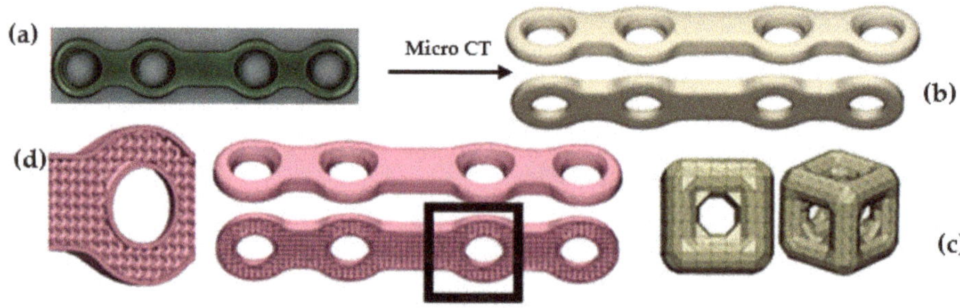

Figure 5. Preparation of a bone fixation plate with an extreme level of porosity and the realistic geometry. (**a**) Conventionally fabricated Ti64 bone fixation plate with standard geometry used as the reference, (**b**) reconstructed CAD through Micro CT, (**c**) porosity cell, and (**d**) porous bone fixation plate.

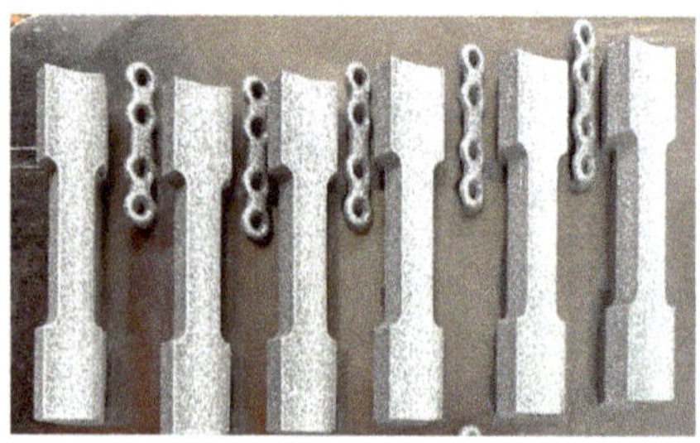

Figure 6. NiTi porous bone fixation plates and tensile samples fabricated via the SLM method.

5.1. *Thermomechanical Behavior*

DSC results, as well as the transformation temperatures of Ni-rich porous bone fixation plates, standard dog-bone samples, and powder particles, are shown and reported in Figure 7 and Table 4, respectively. For the porous plate, the DSC samples were cut from three different regions as shown in Figure 6. Based on the location that the DSC samples were cut from, different transformation temperatures were achieved. The transformation temperatures (TTs) variation can be explained by the nature of the SLM process and the complex shape of the porous plates that resulted in different thermal histories. The effect of different thermal histories during the SLM process can be compared to different heat treatment procedures, which are performed on the final parts. As different heat treatment procedures lead to different TTs, for instance, different thermal histories also can lead to non-homogeneous microstructure [42] as a result, different TTs. This non-homogeneity and variation in TTs variation indicate the necessity of the heat treatment as a post-processing procedure for the as-fabricated samples for creating a homogeneous part. In this study, we did not optimize the heat treatment procedure for the bone fixation plates. That will be studied at a later time.

Table 4. The transformation temperatures (TTs) of powder, dog-bone, and the porous samples.

Sample	M_s (°C)	M_f (°C)	A_s (°C)	A_f (°C)
Powder	10.2	−46.6	−16.9	33.1
Porous #1	12.89	−57.08	−15.82	33.85
Porous #2	0.62	−56.09	−15.53	22.35
Porous #3	2.34	−54.36	−18.72	17.74
Dog-bone	2.14	−60.06	−15.10	26.96

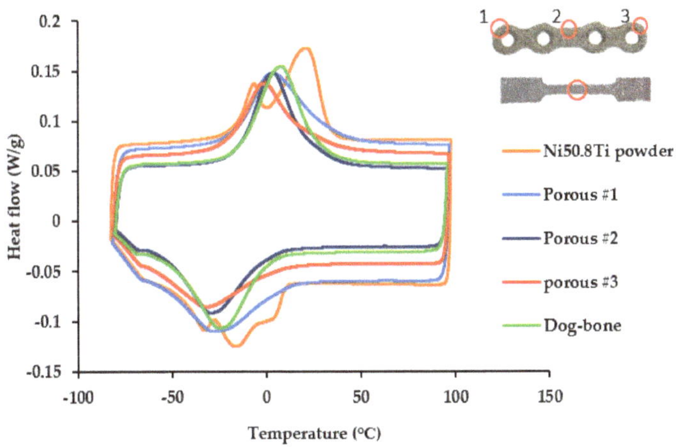

Figure 7. The differential scanning calorimetry (DSC) results of the as-fabricated porous bone fixation plate from three different locations, the dog bone sample, and Ni50.8Ti powder.

To measure the material properties of the fabricated Ni-rich bone fixation plates and updating the FE model, two standard tensile samples were tested, and the results are shown in Figure 8. The first sample was loaded up to the failure with a low strain rate (0.0001 S^{-1}) to maintain the isothermal condition. Based on the stress-strain plot of the first sample, a loading-unloading test on the second sample was designed. The second sample was loaded up to the end of the plateau region (3% strain) and then unloaded to zero loads. The transformation strain of 2.6% (81% of total strain) was achieved for the as-fabricated part under tension. The material showed around 0.6% of irrecoverable strain that could be a result of permanent strain (slip) or the locked-in detwinned martensite and can be recovered by heating [43].

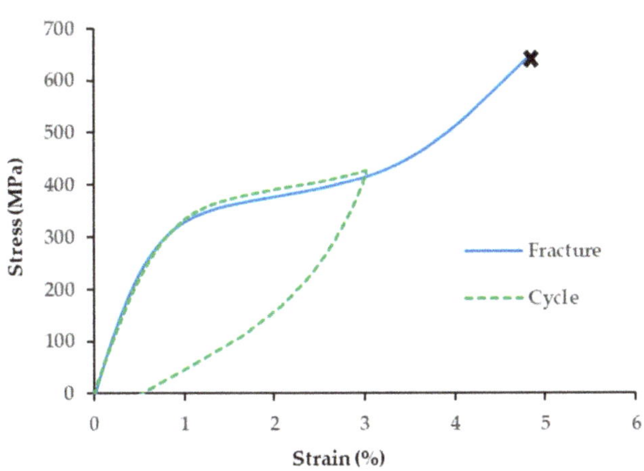

Figure 8. Stress–strain response of SLM dog-bone samples.

The mechanical properties of the as-fabricated material were reported in Table 5. The austenite (EA) and martensite (EM) modulus of elasticity were captured from loading and unloading plots, respectively. Followed by obtaining the material properties from the tensile samples, the Ni-rich bone fixation plate was simulated under tensile loading and unloading. The as-fabricated Ni-rich bone fixation plates were then mechanically tested under the same boundary conditions. Figure 9 shows the simulations as well as mechanical testing of the Ni-rich bone fixation plate.

Table 5. The mechanical properties of the standard tensile sample fabricated via SLM.

E_A (GPa)	E_M (GPa)	Critical Stress Start (MPa)	Critical Stress Finish (MPa)
54	36	350	430

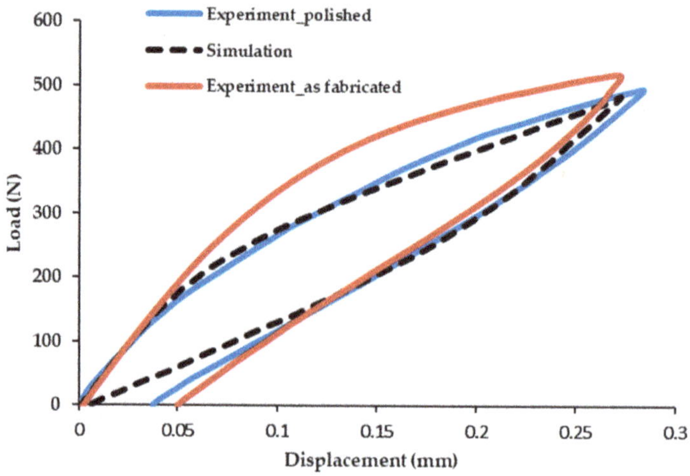

Figure 9. The load–displacement response of as-fabricated porous plate under tensile loading.

After inspection of the fabricated Ni-rich bone fixation plates, we observed a relative geometrical difference between the fabricated part and the CAD model. The size had increased in comparison with the previous case (i.e., the first series of bone fixation plates with larger dimensions). As reported by others [27], as-fabricated SLM fabricated parts could exhibit a geometrical expansion that is related to the laser width, melt pool diameter, and the fact that the part is surrounded by loose powder. The geometrical expansion of SLM-fabricated parts has a more significant effect when the parts include fine details that are close to the laser width. In the case of fabricating Ni-rich bone fixation plates with the extreme geometrical features, the geometrical expansion led to a higher difference with the simulations and the expected behavior. In addition, because of the fine features of the Ni-rich bone fixation plates, a higher amount of un-melted powder was trapped in the pore cells that affect the mechanical response and could affect their biocompatibility. Therefore, we have investigated a chemical polishing method to remove the un-melted powder particles attached to the parts, while not disturbing the fine geometries. This procedure is discussed in the next section. Eventually, the polished bone fixation plate that showed 23% mass loss was mechanically tested, and the load–displacement response is shown in Figure 9.

5.2. Chemical Polishing

Using the selective laser melting process, the complex geometries required for the bone fixation plates can be accurately fabricated. However, the SLM process often results in un-melted powders and other undesirable features that remain fused to the fabricated product. In order to remove the un-melted powder and to add a smooth finish to the part to improve performance (e.g., mechanical, cleaning, and sterilization reliability), a finishing process is required. Due to the complex geometries and fine porous structure of Ni-rich implants, current mechanical finishing processes are not efficient. They may damage the components and cannot reach fine internal features. Therefore, a chemical etching process was opted for. The chemical etching and finishing processes of additively manufactured nickel–titanium components is not well researched. Therefore, a series of tests were undertaken in order to better optimize the etching solution used for the bone fixation plates.

The tests consisted of two different etching solutions and three different times for submerging the components within those solutions (Table 3). The samples were cleaned in water using an ultrasonic bath for 5 min both before and after the chemical finishing process. Both solutions were composed of hydrofluoric acid (HF), nitric acid (HNO_3), and purified distilled water (H_2O). It is known that the HF in the etching solution act as a dissolvent while the HNO_3 behaves as a passivator [44]. The distilled water is used to dilute the solution in order to prevent excessively rapid corrosion/material loss. The samples were submerged for 2 min, 4 min, and 6 min.

Figure 10 shows the results of the chemical etching process on the mass loss of the samples etched the two trialed solutions. Three bone-plate was etched within the etching solution 1 while three others were etched within the solution 2. The mass loss increased proportionally with the time for both solutions. The results show that the second solution with reduced HF and increased HNO_3, removed material at nearly twice the rate of solution 1.

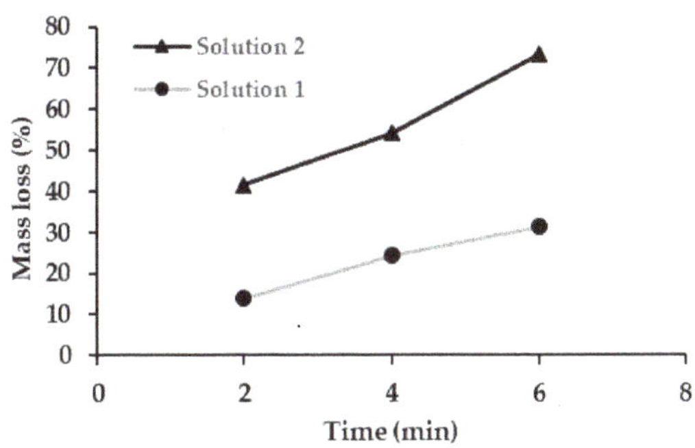

Figure 10. The percentage of mass loss versus time for two different etching solutions.

The effect of the etching procedure on the geometry, as well as the surface finish of the samples, is reflected in the SEM images in Figure 11. As it can be seen in the "As-built" sample, excessive un-melted powder is attached to the surface of the as-built sample and almost entirely covered the pores. S1-2, S1-4, and S1-6 show the SEM images of samples etched with solution 1 with the different durations of 2, 4, and 6 min. The un-melted powder was partially removed on S1-2. S1-4 and S1-6 that were submerged for a long time showed the best result of the chemical polishing. In order to make a more accurate comparison on the effect of chemical polishing, the distance between the pores was measured and compared with the original distance on the CAD file. Based on this comparison, the size and geometry of S1-2 had the most consistency with the original CAD model, and therefore it was reported as the optimized polishing procedure. The polishing by solution 2 for 2 min (S2-2) was almost good but still more than required. While the polishing with solution 2 for 4 and 6 min (S2-4 and S2-6) was too aggressive and caused excessive degradation of the pore structures.

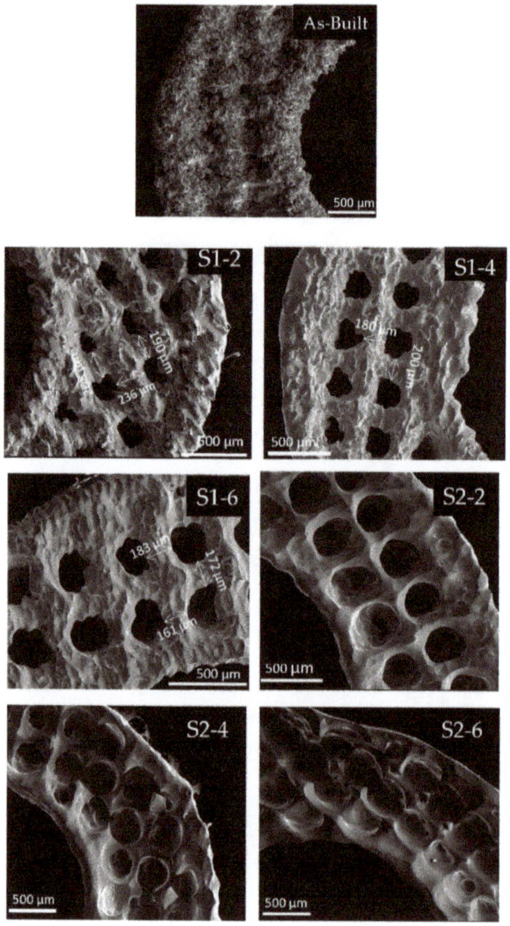

Figure 11. The SEM result of chemically polished porous plates under six different etching conditions.

5.3. *Chemical Composition Analysis*

Understanding the chemical composition of bone fixation plates is critical. Since this is a medical implant, strict controls are required to ensure biocompatibility (i.e., lack of toxicity). Analyzing and accounting for any precipitates or ion formation due to the additive manufacturing process is, therefore, a necessity. The results for our chemical analysis of these bone fixation plates is displayed in Table 6. The elements analyzed were carbon (C), hydrogen (H), oxygen (O), nitrogen (N), nickel (Ni), cobalt (Co), copper (Cu), chromium (Cr), iron (Fe), niobium (Nb), and titanium (Ti). The composition analysis is compared with the allowable values of nickel–titanium shape memory alloys for medical devices and surgical implants, as described in ASTM F 2063 [45]. The results show that the AM Nitinol compositions were within expected parameters with respect to total weight percentage. However, the nitrogen plus oxygen level was slightly above the acceptable ASTM requirement at this stage in our work. The cobalt, copper, chromium, nitrogen, carbon, hydrogen, and niobium were not present in a quantity sufficient to alter the predicted mechanical performance. Ratios of iron and oxygen were also low and do not present an obstacle to the performance and safety of these bone fixation plates.

Table 6. The results of the chemical composition analysis.

Element.	Composition (wt. %)	ASTM F-2063 Composition Requirements (wt. %)
Nickel	56.0	54.5–57.0
Carbon	0.002	<0.050
Cobalt	<0.0005	<0.050
Copper	<0.0005	<0.010
Chromium	<0.0005	<0.010
Hydrogen	0.0004	<0.005
Iron	0.016	<0.050
Niobium	<0.002	<0.025
Oxygen	0.058	-
Nitrogen	<0.005	-
Nitrogen + Oxygen	<0.063	<0.050
Titanium	Balance	Balance

6. Conclusions

The porous stiffness-modulated NiTi bone fixation plates with different levels of stiffness were designed and modeled via a FE analysis. It was shown that specific levels of stiffness could be achieved by introducing engineered levels of porosity to the bulk plate, and the mechanical properties can be captured using the properties of standard coupons. Due to the complex geometry of porous bone fixation plates, the SLM technique was employed to fabricate the NiTi plate. This methodology can be applied to complex geometries and plates as thin as 1.5 mm. The as-fabricated parts were tested mechanically, and the experimental results were in a good agreement with the FE model. After modeling verification and confirming the design methodology, Ni-rich bone fixation plates were fabricated to achieve superelastic behavior. The DSC results of the porous Ni-rich bone fixation plate showed a small variation in TTs of different locations of the plates. Various thermal histories that the parts experienced during the SLM process were the main reason for the variation in TTs. Two tensile samples using Ni-rich powder were fabricated along with Ni-rich bone fixation plates and tested to measure the mechanical properties used for FE modeling. The as-fabricated porous Ni-rich bone-plate fixation was tested and the predicted superelastic behavior was achieved. In order to test methods to remove un-melted powder particles from the surface following AM, the bone-plates were chemically polished under six different conditions. SEM analysis revealed that an etching solution with a composition of HF (10%), HNO_3 (40%), H_2O (50%), and the effective time of 4 min, was the optimal etching solution to remove un-melted powder particles. The chemical composition analysis of our Ni-rich SLM bone fixation plates confirmed that the composition was in a good agreement with the ASTM 2063 requirements.

Author Contributions: Conceptualization, M.E. and D.D.; additive manufacturing, M.N.; simulation and validation, A.J. and K.S.; characterization P.B., K.S., A.J., M.N., G.G. and H.D.; all authors contributed to the writing of initial draft and final draft and reviewed the manuscript. All authors have read and agree to the published version of the manuscript.

Funding: This research was partially funded by a grant from the Third Frontier (Ohio Development Office) of Ohio (grant # TECG20170372).

Acknowledgments: The authors would like to acknowledge the financial support of the Ohio Third Frontier Technology Validation and Start-up Fund.

Conflicts of Interest: The authors declare no conflict of interest.

References

1. Emam, H.A.; Stevens, M.R. Concepts in Bone Reconstruction for Implant Rehabilitation. In *A Textbook of Advanced Oral and Maxillofacial Surgery*; IntechOpen: London, UK, 2013.
2. Jahadakbar, A.; Moghaddam, N.S.; Amerinatanzi, A.; Dean, D.; Karaca, H.; Elahinia, M. Finite element simulation and additive manufacturing of stiffness-matched niti fixation hardware for mandibular reconstruction surgery. *Bioengineering* **2016**, *3*, 36. [CrossRef] [PubMed]
3. Ibrahim, H.; Jahadakbar, A.R.; Dehghan, A.; Moghaddam, N.S.; Amerinatanzi, A.; Elahinia, M. In vitro corrosion assessment of additively manufactured porous NiTi structures for bone fixation applications. *Metals (Basel)* **2018**, *8*, 164. [CrossRef]
4. Moghaddam, N.S.; Jahadakbar, A.; Amerinatanzi, A.; Elahinia, M.; Miller, M.; Dean, D. Metallic fixation of mandibular segmental defects: Graft immobilization and orofacial functional maintenance. *Plast. Reconstr. Surg. Glob. Open* **2016**, *4*, 858. [CrossRef] [PubMed]
5. Moghaddam, N.S.; Jahadakbar, A.; Amerinatanzi, A.; Skoracki, R.; Miller, M.; Dean, D.; Elahinia, M. Fixation release and the bone bandaid: A new bone fixation device paradigm. *Bioengineering* **2017**, *4*, 5. [CrossRef]
6. Jahadakbar, A.; Moghaddam, N.S.; Amerinatanzi, A.; Dean, D.; Elahinia, M. Mechanical Evaluation of the SLM Fabricated, Stiffness-matched, Mandibular Bone Fixation Plates. *Tissue Eng. Part A* **2017**, *23*, S141.
7. Chirani, S.A.; Barati, M.; Saint-Sulpice, L.; Bayati, P.; Nematollahi, M.; Mahtabi, M.; Jahadakbar, A.; Elahinia, M. Self-heating and fatigue of additively manufactured NiTi. In Proceedings of the 2nd International Conference on Simulation for Additive Manufacturing, Pavia, Italy, 11–13 September 2019.
8. Nematollahi, M.; Baghbaderani, K.S.; Amerinatanzi, A.; Zamanian, H.; Elahinia, M. Application of NiTi in Assistive and Rehabilitation Devices: A Review. *Bioengineering* **2019**, *6*, 37. [CrossRef]
9. Morgan, N.B. Medical shape memory alloy applications—The market and its products. *Mater. Sci. Eng. A* **2004**, *378*, 16–23. [CrossRef]
10. Parsai, E.; Elliott, D.; Bayatimalayeri, P.; Elahinia, M. Rectal Retractor, A Novel Device for Sparing Rectum in Radiotherapy Applications of Pelvic Tumors. *Med. Phys.* **2019**, *46*, E199.
11. Parsai, E.I.; Jahadakbar, A.; Lavvafi, H.; Elahinia, M. A novel and innovative device to retract rectum during radiation therapy of pelvic tumors. *J. Appl. Clin. Med. Phys.* **2019**, *20*, 194–199. [CrossRef]
12. Amerinatanzi, A.; Zamanian, H.; Moghaddam, N.S.; Jahadakbar, A.; Elahinia, M. Application of the superelastic NiTi spring in ankle foot orthosis (AFO) to create normal ankle joint behavior. *Bioengineering* **2017**, *4*, 95. [CrossRef]
13. Malayeri, P.B.; Jahadakbar, A.; Elliottt, D.; Elahinia, M.; Parsai, E. A Novel Device to Retract Rectum during Radiotherapy Applications of Pelvic Tumors. *Med. Phys.* **2018**, *45*, E665.
14. Nematollahi, M.; Mehrabi, R.; Callejas, M.A.; Elahinia, H.; Elahinia, M. A two-way architectural actuator using NiTi SE wire and SME spring. *Proc. SPIE* **2018**, *10595*, 105952J9.
15. Andani, M.T.; Haberland, C.; Walker, J.M.; Karamooz, M.; Turabi, A.S.; Saedi, S.; Rahmanian, R.; Karaca, H.; Dean, D.; Kadkhodaei, M. Achieving biocompatible stiffness in NiTi through additive manufacturing. *J. Intell. Mater. Syst. Struct.* **2016**, *27*, 2661–2671. [CrossRef]
16. Saghaian, S.E.; Moghaddam, N.S.; Nematollahi, M.; Saedi, S.; Elahinia, M.; Karaca, H.E. Mechanical and shape memory properties of triply periodic minimal surface (TPMS) NiTi structures fabricated by selective laser melting. *Biol. Eng. Med.* **2018**, *3*. [CrossRef]
17. Dadbakhsh, S.; Speirs, M.; van Humbeeck, J.; Kruth, J.-P. Laser additive manufacturing of bulk and porous shape-memory NiTi alloys: From processes to potential biomedical applications. *MRS Bull.* **2016**, *41*, 765–774. [CrossRef]
18. Tan, C.; Li, S.; Essa, K.; Jamshidi, P.; Zhou, K.; Ma, W.; Attallah, M.M. Laser Powder Bed Fusion of Ti-rich TiNi lattice structures: Process optimisation, geometrical integrity, and phase transformations. *Int. J. Mach. Tools Manuf.* **2019**, *141*, 19–29. [CrossRef]
19. Namatollahi, M.; Jahadakbar, A.; Mahtabi, M.J.; Elahinia, M. Additive manufacturing (AM). In *Metals for Biomedical Devices*; Elsevier: Sawston, UK, 2019; pp. 331–353.
20. Zadpoor, A. Additively manufactured porous metallic biomaterials. *J. Mater. Chem. B* **2019**, *7*, 4088–4117. [CrossRef]
21. Mehrpouya, M.; Gisario, A.; Elahinia, M. Laser welding of NiTi shape memory alloy: A review. *J. Manuf. Process.* **2018**, *31*, 162–186. [CrossRef]

22. Mehrpouya, M.; Dehghanghadikolaei, A.; Fotovvati, B.; Vosooghnia, A.; Emamian, S.S.; Gisario, A. The potential of additive manufacturing in the smart factory industrial 4.0: A review. *Appl. Sci.* **2019**, *9*, 3865. [CrossRef]
23. Rana, M.; Chui, C.H.K.; Wagner, M.; Zimmerer, R.; Rana, M.; Gellrich, N.-C. Increasing the accuracy of orbital reconstruction with selective laser-melted patient-specific implants combined with intraoperative navigation. *J. Oral Maxillofac. Surg.* **2015**, *73*, 1113–1118. [CrossRef]
24. Rotaru, H.; Schumacher, R.; Kim, S.-G.; Dinu, C. Selective laser melted titanium implants: A new technique for the reconstruction of extensive zygomatic complex defects. *Maxillofac. Plast. Reconstr. Surg.* **2015**, *37*, 12. [CrossRef]
25. Yang, W.; Choi, W.S.; Leung, Y.Y.; Curtin, J.P.; Du, R.; Zhang, C.; Chen, X.; Su, Y. Three-dimensional printing of patient-specific surgical plates in head and neck reconstruction: A prospective pilot study. *Oral Oncol.* **2018**, *78*, 31–36. [CrossRef] [PubMed]
26. Elahinia, M.; Moghaddam, N.S.; Andani, M.T.; Amerinatanzi, A.; Bimber, B.A.; Hamilton, R.F. Fabrication of NiTi through additive manufacturing: A review. *Prog. Mater. Sci.* **2016**, *83*, 630–663. [CrossRef]
27. Nematollahi, M.; Toker, G.; Saghaian, S.E.; Salazar, J.; Mahtabi, M.; Benafan, O.; Karaca, H.; Elahinia, M. Additive Manufacturing of Ni-Rich NiTiHf 20: Manufacturability, Composition, Density, and Transformation Behavior. *Shape Mem. Superelasticity* **2019**, *5*, 113–124. [CrossRef]
28. Van Humbeeck, J. Additive Manufacturing of Shape Memory Alloys. *Shape Mem. Superelasticity* **2018**, *4*, 309–312. [CrossRef]
29. Toker, G.P.; Nematollahi, M.; Saghaian, S.E.; Baghbaderani, K.S.; Benafan, O.; Elahinia, M.; Karaca, H.E. Shape memory behavior of NiTiHf alloys fabricated by selective laser melting. *Scr. Mater.* **2020**, *178*, 361–365. [CrossRef]
30. Zhou, Q.; Hayat, M.D.; Chen, G.; Cai, S.; Qu, X.; Tang, H.; Cao, P. Selective electron beam melting of NiTi: Microstructure, phase transformation and mechanical properties. *Mater. Sci. Eng. A* **2019**, *744*, 290–298. [CrossRef]
31. Biffi, C.A.; Bassani, P.; Nematollahi, M.; Moghaddam, N.S.; Amerinatanzi, A.; Mahtabi, M.J.; Elahinia, M.; Tuissi, A. Effect of Ultrasonic Nanocrystal Surface Modification on the Microstructure and Martensitic Transformation of Selective Laser Melted Nitinol. *Materials (Basel)* **2019**, *12*, 3068. [CrossRef]
32. Mehrpouya, M.; Gisario, A.; Rahimzadeh, A.; Nematollahi, M.; Baghbaderani, K.S.; Elahinia, M. A prediction model for finding the optimal laser parameters in additive manufacturing of NiTi shape memory alloy. *Int. J. Adv. Manuf. Technol.* **2019**, *105*, 4691–4699. [CrossRef]
33. Karaji, Z.G.; Speirs, M.; Dadbakhsh, S.; Kruth, J.-P.; Weinans, H.; Zadpoor, A.A.; Yavari, S.A. Additively manufactured and surface biofunctionalized porous nitinol. *ACS Appl. Mater. Interfaces* **2017**, *9*, 1293–1304. [CrossRef]
34. Katona, B.; Bognár, E.; Berta, B.; Nagy, P.; Hirschberg, K. Chemical etching of nitinol stents. *Acta Bioeng. Biomech.* **2013**, *15*, 3–8.
35. Shabalovskaya, S.A.; Anderegg, J.; Laab, F.; Thiel, P.A.; Rondelli, G. Surface Conditions of Nitinol Wires, Tubing, and As-Cast Alloys. The Effect of Chemical Etching, Aging in Boiling Water, and Heat Treatment. *J. Biomed. Mater. Res. Part B Appl. Biomater.* **2003**, *65*, 193–203. [CrossRef]
36. Pelton, R.; Dicello, J.; Miyazaki, S. Optimisation of processing and properties of medical grade Nitinol wire. *Minim. Invasive Ther. Allied Technol.* **2000**, *9*, 107–118. [CrossRef]
37. Mehrabi, R.; Kadkhodaei, M.; Ghaei, A. Numerical implementation of a thermomechanical constitutive model for shape memory alloys using return mapping algorithm and microplane theory. *Adv. Mater. Res.* **2012**, *516*, 351–354. [CrossRef]
38. Mehrabi, R.; Kadkhodaei, M.; Andani, M.T.; Elahinia, M. Microplane modeling of shape memory alloy tubes under tension, torsion, and proportional tension–torsion loading. *J. Intell. Mater. Syst. Struct.* **2015**, *26*, 144–155. [CrossRef]
39. Mehrabi, R.; Kadkhodaei, M. 3D phenomenological constitutive modeling of shape memory alloys based on microplane theory. *Smart Mater. Struct.* **2013**, *22*, 25017. [CrossRef]
40. Andani, M.T.; Alipour, A.; Eshghinejad, A.; Elahinia, M. Modifying the torque–angle behavior of rotary shape memory alloy actuators through axial loading: A semi-analytical study of combined tension–torsion behavior. *J. Intell. Mater. Syst. Struct.* **2013**, *24*, 1524–1535. [CrossRef]

41. Andani, M.T.; Saedi, S.; Turabi, A.S.; Karamooz, M.R.; Haberland, C.; Karaca, H.E.; Elahinia, M. Mechanical and shape memory properties of porous Ni50. 1Ti49. 9 alloys manufactured by selective laser melting. *J. Mech. Behav. Biomed. Mater.* **2017**, *68*, 224–231. [CrossRef] [PubMed]
42. Hamilton, R.F.; Bimber, B.A.; Andani, M.T.; Elahinia, M. Multi-scale shape memory effect recovery in NiTi alloys additive manufactured by selective laser melting and laser directed energy deposition. *J. Mater. Process. Technol.* **2017**, *250*, 55–64. [CrossRef]
43. Saedi, S.; Moghaddam, N.S.; Amerinatanzi, A.; Elahinia, M.; Karaca, H.E. On the effects of selective laser melting process parameters on microstructure and thermomechanical response of Ni-rich NiTi. *Acta Mater.* **2018**, *144*, 552–560. [CrossRef]
44. Simka, W.; Kaczmarek, M.; Baron-Wiecheć, A.; Nawrat, G.; Marciniak, J.; Żak, J. Electropolishing and passivation of NiTi shape memory alloy. *Electrochim. Acta* **2010**, *55*, 2437–2441. [CrossRef]
45. ASTM International. ASTM F2063-18, Standard Specification for Wrought Nickel-Titanium Shape Memory Alloys for Medical Devices and Surgical Implants. 2018. Available online: www.astm.org (accessed on 15 December 2019).

metals

Article

Biodegradable Implantation Material: Mechanical Properties and Surface Corrosion Mechanism of Mg-1Ca-0.5Zr Alloy

Yen-Ting Chen, Fei-Yi Hung * and Jie-Cheng Syu

Department of Materials Science and Engineering, National Cheng Kung University, Tainan 701, Taiwan
* Correspondence: fyhung@mail.ncku.edu.tw; Tel.: +886-6-2757-5756-2950; Fax: +886-6234-6290

Received: 19 June 2019; Accepted: 2 August 2019; Published: 6 August 2019

Abstract: Mg alloy is suitable for biomedical implants as the mechanical properties of Mg are close to those of human bone. Ca is a major element in bone and Zr has a great grain refinement effect. Hence, we developed Mg-1Ca-0.5Zr alloy (XK105) as a biodegradable biomaterial and investigated its mechanical properties and surface corrosion mechanism. The results showed that heat treatment made the secondary phase homogeneous. Tensile tests showed that the heat treatment increased ductility, and that the tensile stress results in the extrusion direction showed better ductility than that in the transverse direction because of the fiber texture and extrusion characteristics. Electrochemistry test results showed that XK105 after heat treatment had a lower corrosion rate than that before heat treatment and that of pure Mg. XK105 after heat treatment formed a calcium phosphate layer after immersion in simulated body fluid; this layer protects Mg from corrosion. Surface roughening treatment increased corrosion because pits on the surface promoted pitting corrosion. This study developed Mg-1Ca-0.5Zr alloy as a biomedical implant material. The results can be used as a reference for the biomedical material industry.

Keywords: Mg-Ca-Zr; mechanical properties; erosion; corrosion

1. Introduction

Common biomedical implant materials include metals, polymers, and ceramics. Metal-based biomedical materials have better mechanical properties than those of polymers and ceramics [1] and are thus much more widely used in practical applications. Common metal-based biomedical materials such as stainless steel [2], Ti alloys [3], and Co-based alloys [4] have elastic moduli that are very different from that of human bone. This leads to a stress shielding effect, which makes new bone tissues lack stimulation during growth and leads to insufficient recovery of bone strength [5]. Another disadvantage of common metal-based biomedical materials is their non-degradability [6]; they must be removed in a second operation after the affected area has recovered.

In view of these issues, Mg alloys have been developed as next-generation biomedical materials [7]. Mg alloys have several advantages as biomaterials. Mg is an essential mineral nutrient, so it has good biosafety [8]. Mg alloys have biomechanical properties similar to those of human bone, which can avoid stress shielding effects and improve the strength of bone tissue after healing [9]. Mg is degradable in vivo, and when used for short-term implants, it does not need to be removed in a second operation [10]. The products produced by the degradation of Mg in vivo can be excreted through the kidney [11]. Mg alloy combines well with surrounding bone tissues after degradation and accelerates the formation of biological calcium phosphate [12]. Mg alloys can stimulate the differentiation and growth of osteoblasts and promote bone tissue production [13].

The concentration of chloride ions in the human body is as high as 150 mmol/L, making the degradation of pure Mg very fast [14]. Such a fast degradation rate may cause slow recovery of the affected area and even lead to tissue necrosis [15]. Adding other metallic elements to form Mg alloys can solve this issue. Mg-Ca alloys have become a focus in the development of Mg-based biomedical materials [16] because Ca is not only harmless to the human body, but also an important component that makes up bone [17]. The Ca ions released during degradation help the healing of bone tissue [18]. Adding Ca to Mg alloys can enhance their mechanical properties through grain refinement and solid solution strengthening [19]. However, if the Ca content is excessive, a large amount of Mg_2Ca will be precipitated, decreasing mechanical properties and corrosion resistance [20]. Mg_2Ca has a lower corrosion potential than that of the Mg base, so it is preferentially corroded as a sacrificial anode and can thus increase the corrosion resistance of the Mg base [21]. However, in excess amounts, Mg_2Ca forms a continuous network structure at the grain boundary, causing cracks [22]. According to previous studies, Mg alloy with 1 wt.% Ca has good mechanical properties and corrosion resistance [23]. Both Zr and Mg are hexagonal close-packed (HCP) structures and the two elements have very similar lattice constants, so adding Zr to Mg alloy can refine grains and improve mechanical properties [24,25]. However, in excess amounts, Zr aggregates into particles larger than 5 μm and reduces the corrosion resistance of Mg alloys [26]. The casting Mg-Ca-Zr alloy has better tensile strength and elongation than Mg-Ca alloy at all Ca content ratios [27]. Heat treatment is important for Mg alloys in application. The temperature of post heat treatment is generally around 380–420 °C for 6–12 h [28,29] to achieve the best mechanical properties. For a novel Mg alloy, the heat treatment conditions should be suggested in this range.

In this study, the Mg alloy Mg-1Ca-0.5Zr (98.5 wt.% Mg, 1 wt.% Ca, and 0.5 wt.% Zr) is subjected to heat treatment to homogenize and stabilize its material properties. The heat treatment condition was 400 °C for 8 h at atmosphere which followed the common condition of Mg-Ca alloy [28]. A series of experiments were conducted to evaluate the feasibility of Mg-1Ca-0.5Zr alloy as a biomedical implant material. The results can be used as a reference for the biomedical material industry.

2. Material and Methods

Mg-1Ca-0.5Zr was produced in a smelting furnace. A 99.9 wt.% pure Mg block was placed in a stainless-steel bowl and heated to 700 °C until the Mg block had completely melted into a liquid state. Then, Mg-20Ca and Mg-33Zr blocks were placed into the molten Mg. After the alloy blocks had melted completely, the molten alloy was poured into a mold and allowed to cool in air to room temperature. For all casting, argon was used as a protective gas. The casting product, Mg-1Ca-0.5Zr alloy, is denoted as XK105 following the ASTM (American Society for Testing and Materials) naming convention. For experiments, we processed XK105 into an extruded sheet, denoted as XK105-F. To enhance the mechanical properties, XK105 was subjected to heat treatment at 400 °C for 8 h in an air furnace. To decrease the formation of the crystalline phase, the alloy was water-quenched after heat treatment. The heat-treated XK105 is denoted XK105-H.

For microstructural observation, epoxy was used to cold-mount and embed XK105-F and XK105-H. The samples were then polished using 1 and 0.3 μm aluminum oxide powder. After polishing, the samples were corroded using a solution of 4.2 g of picric acid, 10 mL of acetic acid, and 90 mL of ethanol. The microstructure was observed using light microscopy (LM) (ZEISS, Oberkochen, Germany) and scanning electron microscopy (SEM) (HITACHI, Tokyo, Japan). Energy-dispersive X-ray spectrometry (EDS) was used for elemental composition semi-quantitative analysis.

This study used tensile tests to determine the tensile strength of the extruded profiles of XK105 and the mechanical properties after heat treatment. Three specimens of XK105-F were taken along the extrusion direction (ED) and the transverse direction (TD). After the samples were cut to an appropriate size, they were punched into a fixed size using a punch press. A stretching test was conducted with an initial strain rate of 8.3×10^{-4} s^{-1} at room temperature. We used SEM and EDS to investigate the stretch-breaking behavior.

An erosion wear test was conducted using XK105-F and XK105-H. A particle spray tester (Chun Yen Testing Machines Co., Taichung, Taiwan) was used for erosion wear and surface roughening.

The environmental conditions in the erosion wear test were room temperature and a relative humidity of 50% to 70%. The erosion particles were Al_2O_3 (average diameter: 150 μm) and SiO_2 (average diameter: 295 μm), and the erosion angle was 15° to 90°. The particle rate of erosion was 1.2 g/s, and the pressure was set to 3 kg/cm^2 (0.29 MPa). The average jet rate of Al_2O_3 was calculated as 73 m/s, and that of SiO_2 was 66 m/s. Surface roughening treatment was conducted at erosion angles of 15°, 45°, and 90°. After surface roughening, the samples were observed using LM and SEM.

A potentiodynamic polarization test was conducted using pure Mg, XK105-F, and XK105-H samples. In the test, a three-electrode system was used to obtain the polarization curve. The reference electrode was a saturated calomel electrode (SCE), the salt bridge was saturated KCl solution, and the auxiliary electrode was a platinum electrode. The specimens were embedded in epoxy resin and then carefully ground. The employed electrolyte was a revised simulated body fluid (R-SBF) solution [30], which has ion composition and concentration similar to those of human blood plasma. The scan rate in the polarization test was 1 mV/s, from −2.2 to −1.2 V, and a constant temperature of 37 °C was used. The polarization curve has a linear Tafel region around the corrosion potential. The corrosion current can be obtained from the anodization curves using the extrapolation method.

The immersion test was used to evaluate the corrosion rate. The corrosion rate was calculated as:

$$\text{Corrosion rate} = \frac{m_1 - m_0}{A \cdot T}$$

where m_1 is the mass after immersion, m_0 is the mass before immersion, A is the total surface area of the sample, and T is the immersion time. The immersion solution was R-SBF. XK105-H and surface-roughened XK105-H sampled were immersed in R-SBF for 1, 3, and 7 days at a constant temperature of 37 °C.

3. Results and Discussion

3.1. Metallographic Analysis

Figure 1a,b respectively show metallographic images of XK105-F and XK105-H observed using LM. Crystal phases appear along ED for XK105-F. The crystal phases of XK105-H were not along ED. This indicates that heat treatment made the crystal phases dissolve into the matrix, homogenizing the material. The grains were subjected to dynamic recrystallization during extrusion, and thus isometric crystals formed. The calculated average grain sizes of XK105-F and XK105-H were 14.2 and 17.1 μm, respectively.

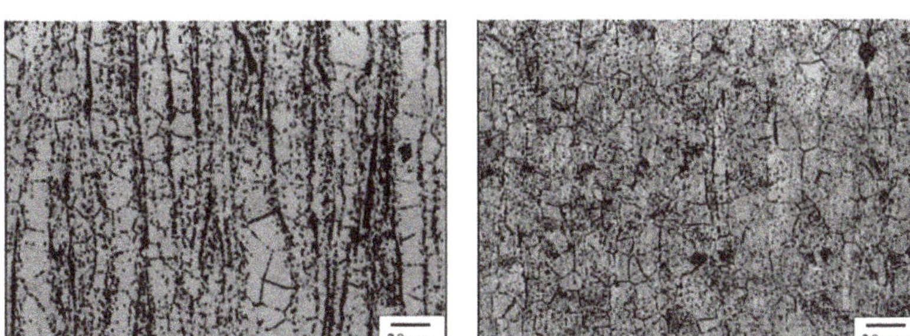

(a) (b)

Figure 1. Light microscopy (LM) microstructural images. (**a**) XK105-F and (**b**) XK105-H.

SEM observation was conducted to verify the secondary phases of XK105. A Ca-rich crystal phase can be seen in Figure 2a; the phase was identified as Mg_2Ca [31]. Most of the Mg_2Ca phase was along

ED. A Zr-rich phase can be seen in Figure 2b; the phase was identified as pure Zr particles which accumulated at grains [32].

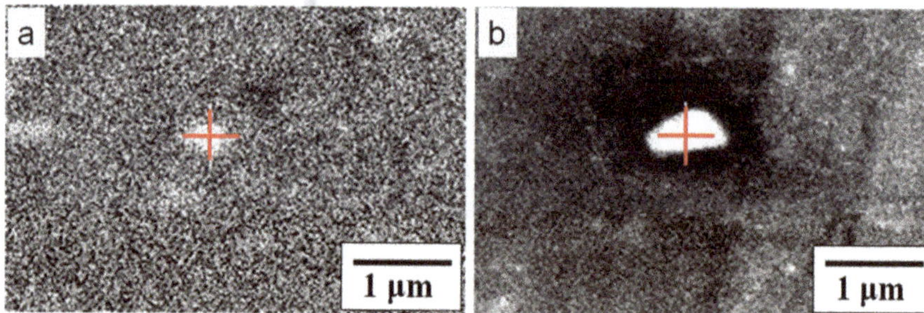

Element	Mg	Ca	Zr
Concentration (at. %)	93.1	6.8	0.1

Element	Mg	Ca	Zr
Concentration (at. %)	20.4	1.3	78.3

Figure 2. Secondary phase analysis of XK105-F using scanning electron microscopy (SEM) and energy-dispersive X-ray spectrometry (EDS). (**a**) Mg_2Ca and (**b**) Zr particles.

3.2. *Tensile Properties and Fracture Morphology*

Figure 3 and Figure 4 respectively show the tensile stress-strain curve and tensile stress of XK105. The tensile stress of XK105-F along TD is about 220 MPa. According to reference [33], the tensile stress along TD of pure Mg was 173 MPa, and which of Mg-0.1Ca was 180 MPa. The tensile stress of human bone is about 150 MPa. XK105 thus has sufficient tensile strength to apply to stress load in the human body. The yield stress of XK105-F along TD was about 180 MPa, which of pure Mg was 103 MPa and which of Mg-0.1Ca was 92 MPa [33]. The results showed the adding Zr can improve the tensile strength effectively. Figure 5 shows the elongation of XK105. The elongation was about 4–8% and was caused by the addition of Ca [34]. Elongation increased after heat treatment because of the homogenization of textures and the solid solution of the secondary phases. Figure 6 shows the Young's modulus of XK105. The value is about 35–55 GPa; it increased after heat treatment because the Ca atoms dissolved into the matrix and changed the lattice. Although the Young's modulus of XK105 is much higher than that of human bone (3–20 GPa), it has greater similarity compared to other medical implant materials. The tensile strength along TD is higher than that along ED. The XRD pattern (Figure 7) shows that the mean peak along ED was $(10\bar{1}1)$. This is due to the HCP crystals slipping along ED, which led to a small angle between the (0001) plane and ED. When the tensile strength transferred toward ED, the dislocations more easily slipped compared to that toward TD. Therefore, the elongation of XK105 along ED was higher but the tensile stress was lower compared to those along TD.

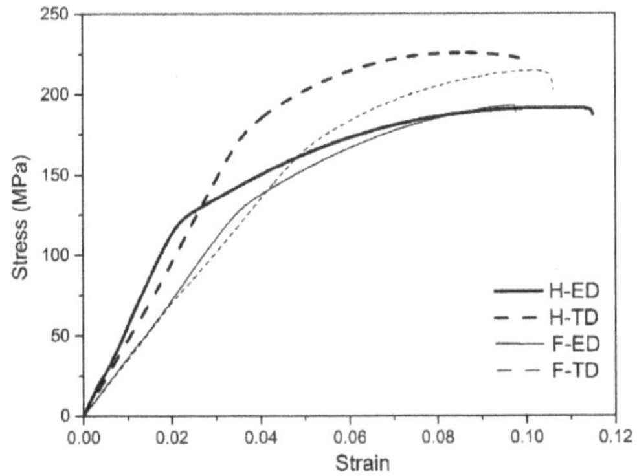

Figure 3. Tensile stress-strain curves of XK105. (H-ED: extrusion direction of XK105-H; H-TD: transverse direction of XK105-H; F-ED: extrusion of XK105-F; F-TD: transverse of XK105-F).

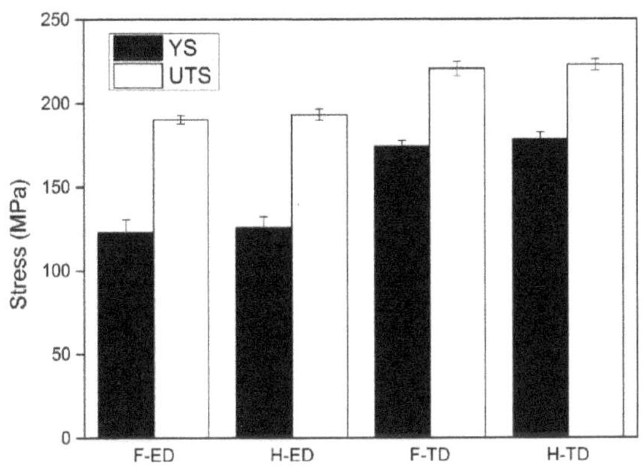

Figure 4. Tensile stress analysis of XK105 (YS: yield stress; UTS: ultimate tensile stress).

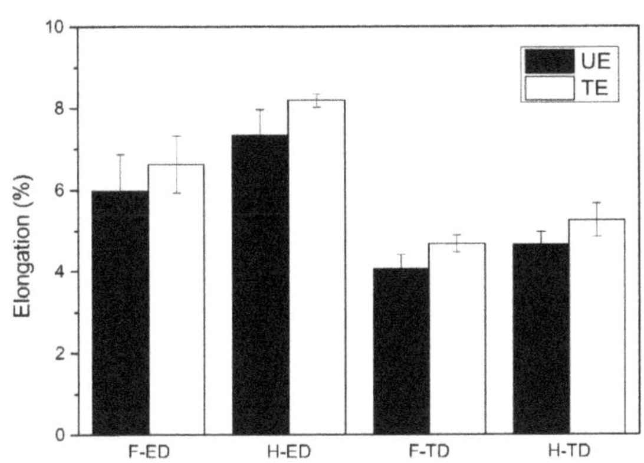

Figure 5. Elongation analysis of XK105 (UE: uniform elongation; TE: total elongation).

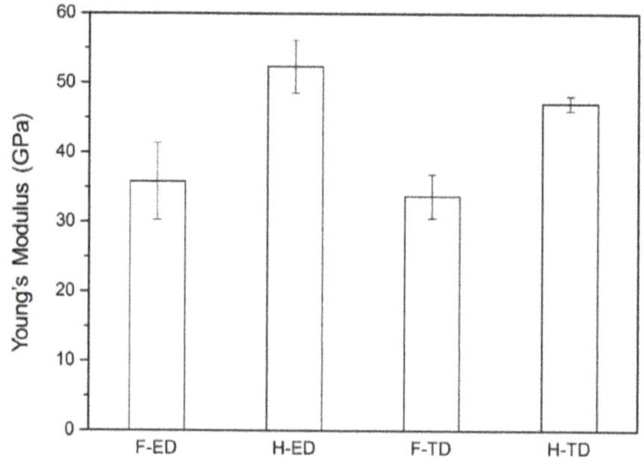

Figure 6. Young's modulus of XK105. (H-ED: extrusion direction of XK105-H; H-TD: transverse direction of XK105-H; F-ED: extrusion of XK105-F; F-TD: transverse of XK105-F).

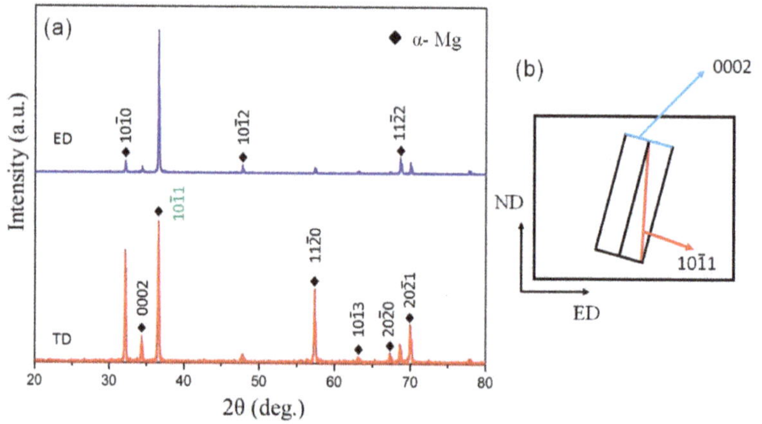

Figure 7. Texture analysis of XK105-H. (**a**) XRD patterns of XK105-H on ED and TD planes and (**b**) illustration of dislocation slip.

Figure 8a,b shows the tensile fracture morphology of XK105-H on the ED plane. Dimples can be observed on the fracture surface. Mg_2Ca, which did not dissolve into the matrix, appears inside the dimples. Figure 8c,d shows the tensile fracture morphology of XK105-H on the TD plane. Thin wall structures can be observed. These structures are cleavages of brittle failure.

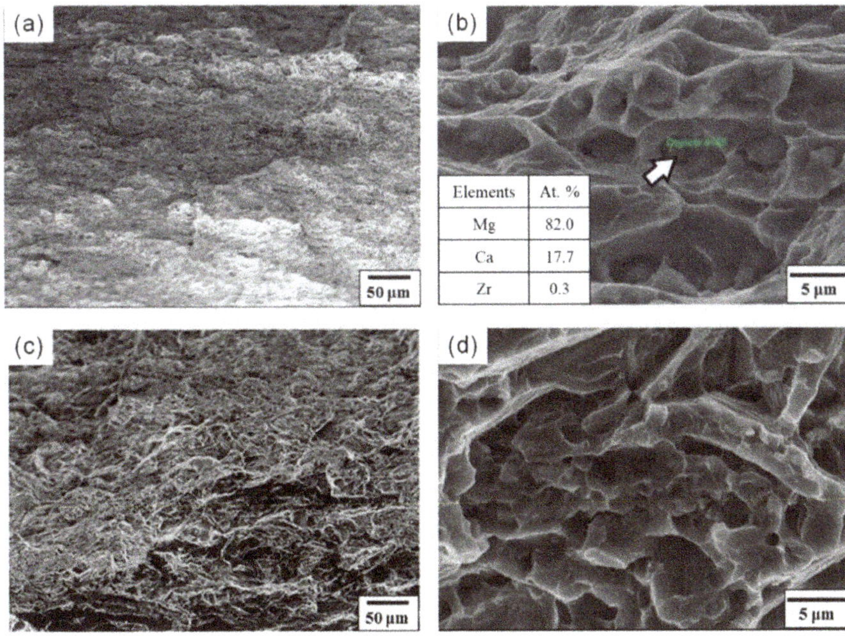

Figure 8. Tensile fracture morphology of XK105-H observed by SEM. Fracture surface on (**a**,**b**) ED plane and (**c**,**d**) TD plane.

3.3. Fracture Toughness

Medical implants easily deform when subjected to a stress impact. The fracture toughness of XK105 is thus an important property. Figure 9 shows the fracture toughness of XK105. The fracture toughness was lower when the stress was applied toward TD. There is a non-plane strain of XK105, and it showed that the thickness of material was not thick enough to fracture as a plane strain. LM observations indicate that the fracture cracks had more bending along ED than TD, as shown in Figure 10a,c. The cracks thus required more energy to spread, so the toughness of the sample along ED was higher than that along TD. Figure 10b,d show the fracture surface of XK105-H. A smoother fracture surface, which resulted from brittle fracture on the TD plane, can be observed. Therefore, toughness on the TD plane was lower than that on the ED plane. Figure 11 shows the secondary surface morphology and fracture mechanism. The end of the crack on the ED plane has a branch morphology; the branching of that on the TD plane is not obvious. This was caused by the brittle secondary phase easily cracking when subjected to stress; the cracks spread along TD while tensile stress was applied along ED. The density of the secondary phase was thus lower on the crack-spreading path. Therefore, the cracks on the ED plane easily branched and had much longer spreading paths than those of cracks on the TD plane.

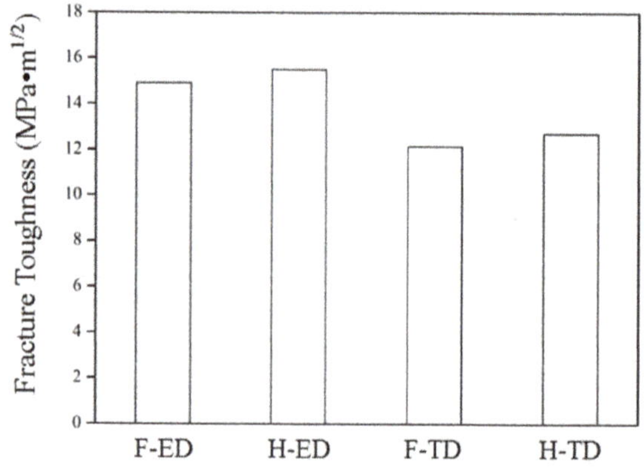

Figure 9. Fracture toughness results for XK105. (H-ED: extrusion direction of XK105-H; H-TD: transverse direction of XK105-H; F-ED: extrusion of XK105-F; F-TD: transverse of XK105-F).

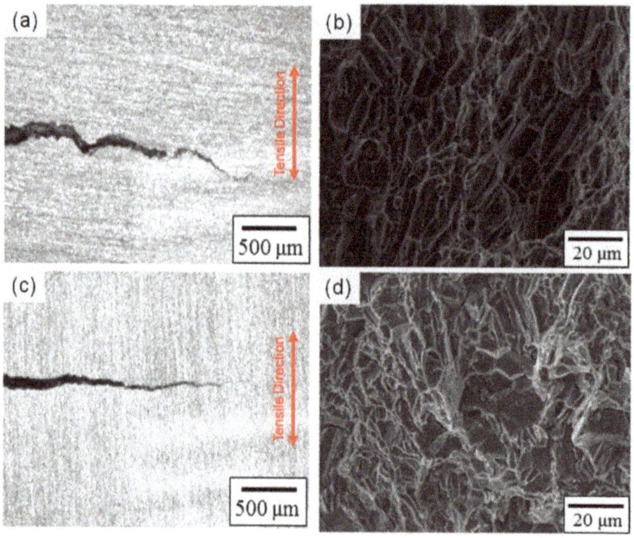

Figure 10. Crack morphology of XK105-H after fracture toughness test. (**a**) LM image on ED plane, (**b**) SEM image on ED plane, (**c**) LM image on TD plane, and (**d**) SEM image on TD plane.

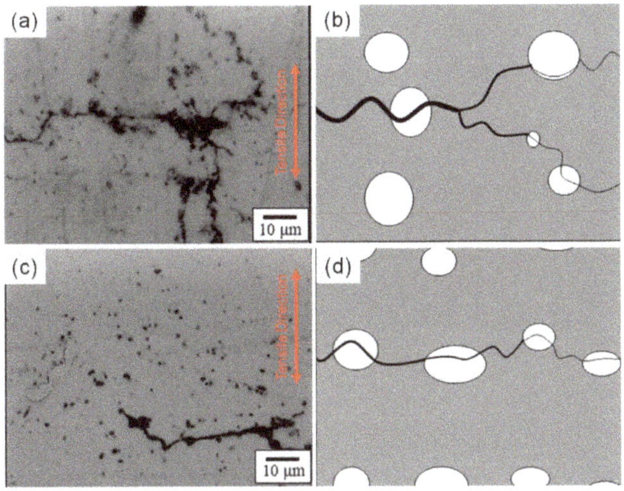

Figure 11. Fracture mechanism of XK105-H. (**a**) LM secondary surface morphology on ED plane, (**b**) fracture mechanism illustration for ED plane, (**c**) LM secondary surface morphology on TD plane, and (**d**) fracture mechanism illustration for TD plane.

3.4. Erosion Properties and Surface Roughness

Figure 12 shows the relationship between erosion rate and erosion angle. For both Al_2O_3 or SiO_2, the angle for the highest erosion rate changed from 45° to 30° after heat treatment. This shows that XK105 had lower hardness and higher ductility after heat treatment. The erosion rate obtained with Al_2O_3 particles was higher than that obtained with SiO_2 due to the higher hardness of Al_2O_3.

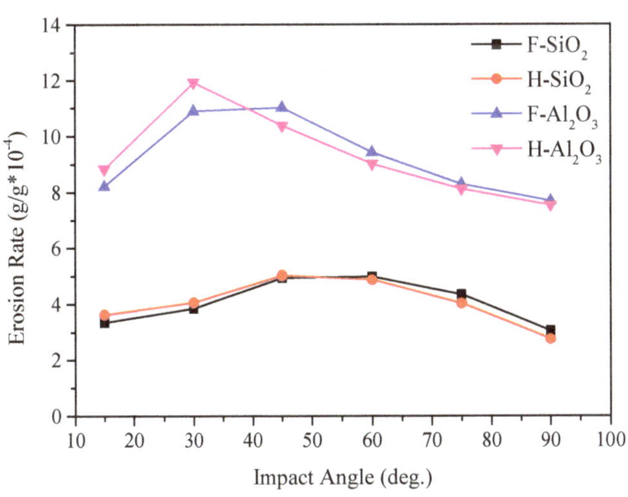

Figure 12. Erosion rate of XK105-F and XK105-H.

Figure 13 shows the surface roughness after the erosion test. The roughness obtained with Al_2O_3 and SiO_2 was similar. The highest and lowest roughness was obtained with erosion angles of 90° and 15°, respectively. The measurement of these erosion properties could be applied to later experiments to clarify the relationship between anti-corrosion ability and surface roughness. This will have a benefit of increasing the bioreactivity for medical applications.

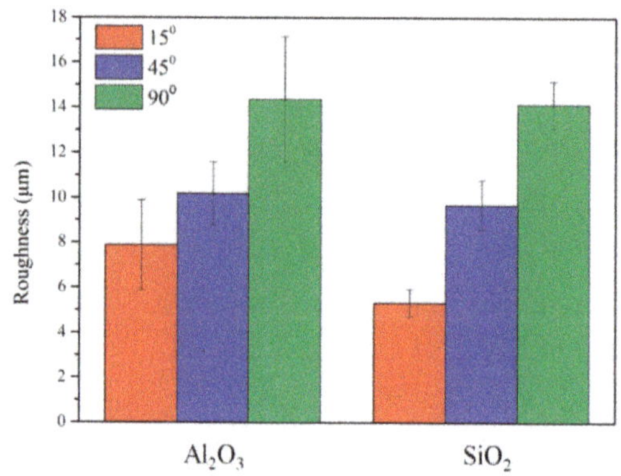

Figure 13. Surface roughness of XK105-H after erosion test.

3.5. Potentiodynamic Polarization Test

The potentiodynamic polarization test was used to analyze the corrosion rate of XK105 in simulated body fluid. Surface roughening treatment can effectively increase cell adhesion [35]. Therefore, in this study, surface roughen treatment using the particle erosion wear method was applied. The potentiodynamic polarization test was conducted to determine anti-corrosion ability. The electrochemical properties of XK105 calculated from the potentiodynamic polarization test results are shown in Table 1. The results show that heat treatment improved the anti-corrosion property of XK105, making it better than that of pure Mg. This was due to Mg_2Ca being dissolved into the matrix during heat treatment, and the effect of galvanic corrosion becoming inactive, which reduced the corrosion current and potential. The potentiodynamic polarization curve is shown in Figure 14. The reduction rate of XK105-F is lower than that of pure Mg. XK105-H had the lowest reduction rate. The low reduction rate represents the low formation rate of H_2, which is harmless to the human body. The curves of XK105-F and XK105-H show passivation at a corrosion potential of less than −1.5 V. Passivation was caused by the added Ca, which formed a stable passivation layer to restrain the corrosion compared to pure Mg [36].

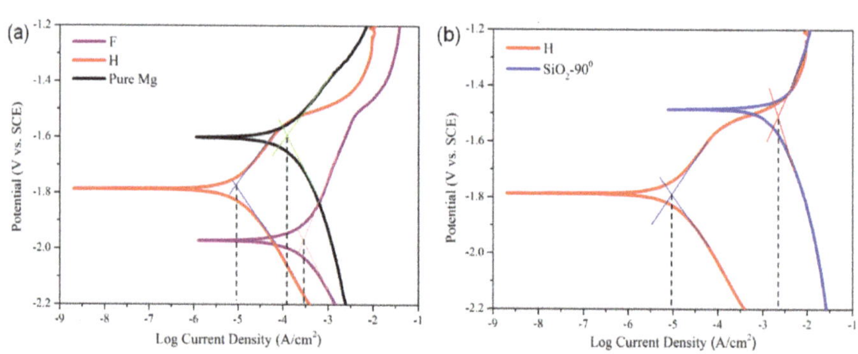

Figure 14. Potentiodynamic polarization curve of XK105. (**a**) Comparison of XK105-F, XK105-H, and pure Mg and (**b**) comparison of XK105-H before and after erosion test by SiO_2 particles with erosion angle of 90°.

Table 1. Electrochemical properties of XK105 and pure Mg.

Sample	Pure Mg	XK105-F	XK105-H
E_{corr} (V)	−1.601	−1.971	−1.786
I_{corr} (A)	3.69×10^{-5}	1.13×10^{-4}	3.85×10^{-6}

For surface roughening, 90° SiO_2 erosion created higher roughness. The electrochemical properties of surface-roughened XK105 are shown in Table 2. The corrosion current was higher after surface roughening treatment. The surface roughening increased the surface area of the material, increasing the corrosion rate [37]. The surface-roughened XK105 did not show passivation around −1.5 V in the potentiodynamic polarization curve because the rough surface made the passivation layer easily peel off after formation.

Table 2. Electrochemical properties of heat treated and surface-roughened XK105.

Sample	XK105-H	SiO_2-90°
E_{corr} (V)	−1.786	−1.485
I_{corr} (A)	3.85×10^{-6}	5.85×10^{-4}

3.6. Immersion Test

To understand the surface corrosion mechanism and quantify the corrosion rate of materials, this study subjected XK105-H and surface-roughened XK105 to an immersion test. The results of the immersion test are shown in Figure 15. The corrosion rate after surface roughening treatment increased greatly, and the corrosion rate of XK105-H was lower than that of pure Mg. The corrosion rate of XK105-H after three days decreased and became stable. This was due to the formation of a surface passivation layer, which restrained the corrosion effect. The phenomenon in which the corrosion rate decreased after three days of surface-roughened XK105 was not obvious relatively, because it did not have a passivation layer.

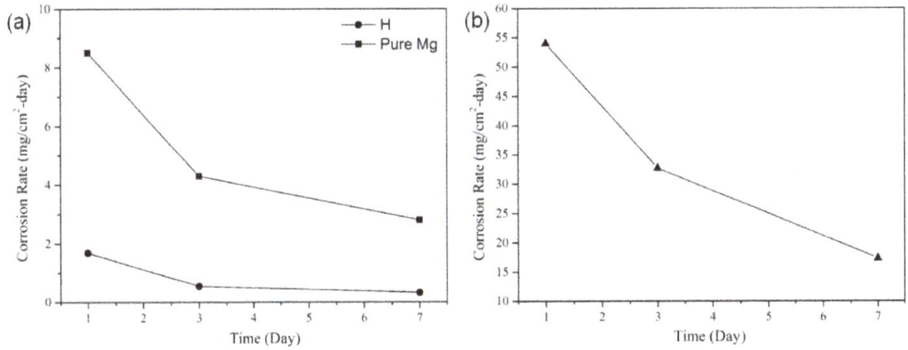

Figure 15. Corrosion rate in immersion test (**a**) XK105-H and pure Mg and (**b**) surface-roughened XK105-H.

Figure 16 shows the secondary surface of XK105-H and surface-roughened XK105 during the immersion test. XK105-H maintained the integrity of its macroscopic morphology, whereas the surface-roughened XK105 showed pitting corrosion after one day of immersion. The main corrosion mechanism of Mg alloy was pitting corrosion when immersed in R-SBF [30]. The surface roughening treatment created a lot of pits on the surface of the material. These pits promoted pitting corrosion, and thus weight loss due to corrosion was observed on the 7th day of immersion.

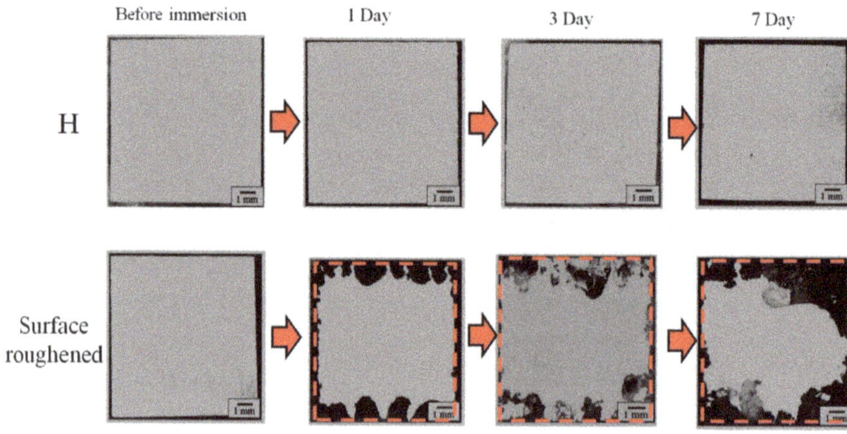

Figure 16. Cross-sectional observation of roughened surface during immersion test.

In Figure 17a, a calcium phosphate layer can be observed on the XK105-H surface after immersion. This layer protected the material against corrosion. Calcium phosphate is the main component of human bone, so this layer could help heal the affected area [38]. In contrast, there was a thick cover layer on the surface of the surface-roughened XK105, as shown in Figure 17b. The cover compound was Mg hydroxide which does not have a healing effect. The formation of calcium phosphate requires a suitable hydroxide ion group to react with [39]. The degradation rate of surface- roughened XK105 was too fast for calcium phosphate to form; instead, Mg hydroxide formed.

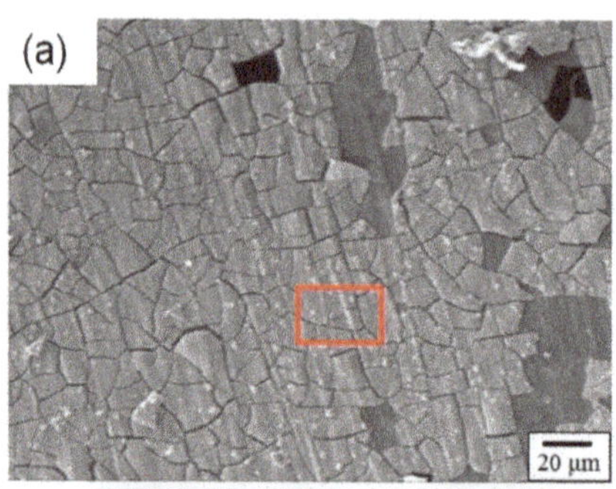

Elements	at. %
Mg	29.7
O	35.5
Ca	15.9
P	18.9

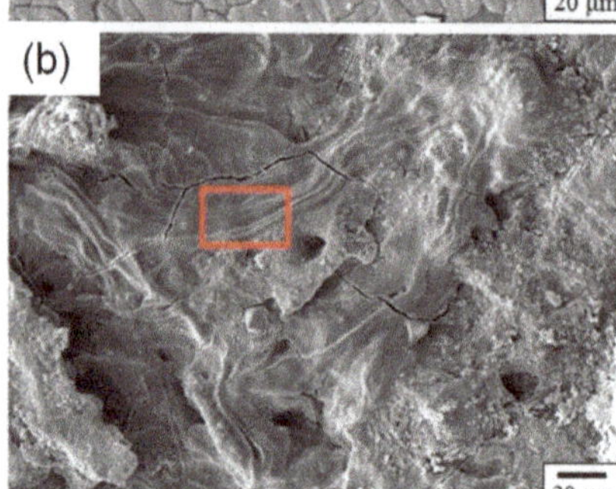

Elements	at. %
Mg	46.3
O	48.2
Ca	1.3
P	4.2

Figure 17. SEM images and EDS analysis of immersed surface. (**a**) XK105-H and (**b**) surface-roughened XK105-H.

To summarize the above results, Mg-1Ca-0.5Zr biomedical alloy has a suitable degradation rate and good mechanical strength. The surface roughness can help control the degradation rate. Mg-1Ca-0.5Zr can thus be applied for biomedical implants. A summary is shown in Figure 18.

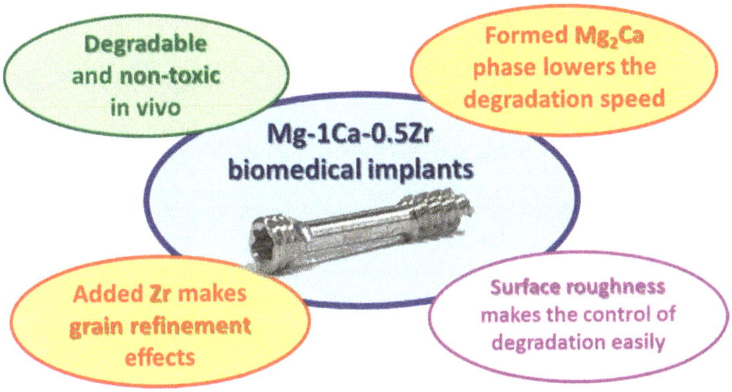

Figure 18. Summary of Mg-1Ca-0.5Zr biomedical implants.

4. Conclusions

(1) The brittle Mg_2Ca phase of the XK105 solid dissolved into the matrix after heat treatment. This made the texture uniform and increased both elongation and fracture toughness. The Young's modulus of XK105 increased after heat treatment because of the solid solution of Ca atoms.

(2) The TD plane of XK105 had higher strength but lower elongation than those of the ED plane because of the texture effects in which the grain slip was hard to start up on the ED plane. The fracture toughness along ED was better than that along TD because the crack-spreading direction was parallel to the arrangement direction of the secondary phase.

(3) The heat treatment of XK105 reduced galvanic corrosion due to the solid solution, and further improved anti-corrosion ability. The surface-roughened XK105 had a larger surface area, which enhanced the corrosion reaction.

(4) A calcium phosphate layer formed during corrosion on the XK105-H surface, significantly reducing the corrosion rate. The surface-roughened XK105 formed magnesium hydroxide instead because of its faster corrosion rate. The magnesium hydroxide did not provide a protective effect.

(5) This study developed a biomedical implant material, Mg-1Ca-0.5Zr alloy. The results can be used as a reference for the biomedical material industry.

Author Contributions: Y.-T.C.: conceived and designed the analysis, collected the data, performed the analysis, wrote the paper. F.-Y.H.: conceived and designed the analysis, contributed data and analysis tools, performed the analysis, revised the paper. J.-C.S.: collected the data, performed the analysis, wrote the paper.

Funding: The authors are grateful to the Instrument Center of National Cheng Kung University and the Ministry of Science and Technology of the Republic of China (MOST 107-2221-E-006-012-MY2) for their financial support.

Conflicts of Interest: The authors declare no conflicts of interest.

References

1. Staiger, M.P.; Pietak, A.M.; Huadmai, J.; Dias, G. Magnesium and its alloys as orthopedic biomaterials: A review. *Biomaterials* **2006**, *27*, 1728–1734. [CrossRef] [PubMed]
2. Dikici, B.; Topuz, M. Production of annealed cold-Sprayed 316L stainless steel coatings for biomedical applications and their in-vitro corrosion response. *Prot. Met. Phys. Chem. Surf.* **2018**, *54*, 333–339. [CrossRef]
3. Yadroitsev, I.; Karkhmalev, P.; Yadroitsava, I.; Du Plessis, A. Qualification of Ti6Al4V ELI alloy produced by laser powder bed fusion for biomedical applications. *JOM* **2018**, *70*, 372–377. [CrossRef]

4. Chen, Q.Z.; Thouas, G.A. Metallic implant biomaterials. *Mater. Sci. Eng. R Rep.* **2015**, *87*, 1–57. [CrossRef]
5. Chien, C.; Liao, T.Y.; Hong, T.F.; Kuo, T.Y.; Chang, C.H.; Yeh, M.L.; Lee, T.M. Surface microstructure and bioactivity of hydroxyapatite and fluorapatite coatings deposited on Ti-6Al-4V substrates using Nd-YAG Laser. *J. Med. Biol. Eng.* **2014**, *34*, 109–115. [CrossRef]
6. Hench, L.L.; Polak, J.M. Third-generation biomedical materials. *Science* **2002**, *295*, 1014–1017. [CrossRef] [PubMed]
7. Kojima, Y. Platform science and technology for advanced magnesium alloys. *Mater. Sci. Forum* **2000**, *350*, 3–18. [CrossRef]
8. Musso, C.G. Magnesium metabolism in health and disease. *Int. Urol. Nephrol.* **2009**, *41*, 357–362. [CrossRef]
9. Nagels, J.; Stokdijk, M.; Rozing, P.M. Stress shielding and bone resorption in shoulder. *J. Shoulder Elb. Surg.* **2003**, *12*, 35–39. [CrossRef]
10. Jurgen, V. Magnesium: Nutrition and metabolism. *Mol. Asp. Med.* **2003**, *24*, 27–37. [CrossRef]
11. Lin, D.J.; Hung, F.Y.; Jakfar, S.; Yeh, M.L. Tailored coating chemistry and interfacial properties for construction of bioactive ceramic coatings on magnesium biomaterial. *Mater. Des.* **2016**, *89*, 235–244. [CrossRef]
12. Witte, F.; Kaese, V.; Haferkamp, H.; Switzer, E.; Meyer-Lindenberg, A.; Wirth, C.J.; Windhagen, H. In vivo corrosion of four magnesium alloys and the associated bone response. *Biomaterials* **2005**, *26*, 3557–3563. [CrossRef] [PubMed]
13. Pietak, A.; Mahoney, P.; Dias, G.J.; Staiger, M.P. Bone-like matrix formation on magnesium and magnesium alloys. *J. Mater. Sci. Mater. Med.* **2008**, *19*, 407–415. [CrossRef] [PubMed]
14. Witte, F.; Hort, N.; Vogt, C.; Cohen, S.; Kainer, K.U.; Willumeit, R.; Feyerabend, F. Degradable biomaterials based on magnesium corrosion. *Curr. Opin. Solid State Mater. Sci.* **2008**, *12*, 63–72. [CrossRef]
15. Hong, Y.S.; Yang, K.; Zhang, G.D.; Huang, J.J.; Hao, Y.Q.; Ai, H.J. The role of Bone induction of a biodegradable magnesium alloy. *Acta Metall. Sin.* **2008**, *44*, 1035–1041.
16. Radha, R.; Sreekanth, D. Insight of magnesium alloys and composites for orthopedic implant applications—A review. *J. Magnes. Alloy.* **2017**, *5*, 286–312. [CrossRef]
17. Harandi, S.E.; Idris, M.H.; Jafari, H. Effect of forging process on microstructure, mechanical and corrosion properties of biodegradable Mg–1Ca alloy. *Mater. Des.* **2011**, *32*, 2596–2603. [CrossRef]
18. Du, H.; Wei, Z.J.; Liu, X.W.; Zhang, E.L. Effects of Zn on the microstructure, mechanical property and bio-corrosion property of Mg–3Ca alloys for biomedical application. *Mater. Chem. Phys.* **2011**, *125*, 568–575. [CrossRef]
19. Kannan, M.B.; Raman, R.K.S. In vitro degradation and mechanical integrity of calcium-containing magnesium alloys in modified-simulated body fluid. *Biomaterials* **2008**, *29*, 2306–2314. [CrossRef]
20. Lin, D.J.; Hung, F.Y.; Lui, T.S.; Yeh, M.L. Heat treatment mechanism and biodegradable characteristics of ZAX1330 Mg alloy. *Mater. Sci. Eng. C* **2015**, *51*, 300–308. [CrossRef]
21. Jeong, Y.S.; Kim, W.J. Enhancement of mechanical properties and corrosion resistance of Mg–Ca alloys through microstructural refinement by indirect extrusion. *Corros. Sci.* **2014**, *82*, 392–403. [CrossRef]
22. Kim, W.C.; Kim, J.G.; Lee, J.Y.; Seok, H.K. Influence of Ca on the corrosion properties of magnesium for biomaterials. *Mater. Lett.* **2008**, *62*, 4146–4148. [CrossRef]
23. Zheng, Y.F.; Gu, X.N.; Xi, Y.L.; Chai, D.L. In vitro degradation and cytotoxicity of Mg/Ca composites produced by powder metallurgy. *Acta Biomater.* **2010**, *6*, 1783–1791. [CrossRef] [PubMed]
24. Qian, M.; StJohn, D.H.; Frost, M.T. Heterogeneous nuclei size in magnesium–zirconium alloys. *Scr. Mater.* **2004**, *50*, 1115–1119. [CrossRef]
25. Liu, G.L. The electronic structure of the microstructure of Mg-Zr alloys. *Acta Phys. Sin.* **2008**, *57*, 1043–1047.
26. Zhou, Y.L.; Li, Y.; Luo, D.M.; Wen, C.; Hodgson, P. Microstructures, mechanical properties and in vitro corrosion behaviour of biodegradable Mg–Zr–Ca alloys. *J. Mater. Sci.* **2012**, *48*, 1632–1639. [CrossRef]
27. Tamura, Y.; Sugimoto, Y.; Soda, H. Structure and mechanical properties of Mg–Ca and Mg–Ca–Zr alloys. *J. Jpn. Inst. Light Met.* **2013**, *63*, 279–285. [CrossRef]
28. Zhong, C.; Liu, F.; Wu, Y.T.; Le, J.J.; Liu, L.; He, M.F.; Zhu, J.C.; Hu, W.B. Protective diffusion coatings on magnesium alloys: A review of recent developments. *J. Alloys Compd.* **2012**, *520*, 11–21. [CrossRef]
29. Ali, Y.H.; Qiu, D.; Jiang, B.; Pan, F.S.; Zhang, M.X. Current research progress in grain refinement of cast magnesium alloys: A review article. *J. Alloys Compd.* **2015**, *619*, 639–651. [CrossRef]
30. Oyane, A.; Kim, H.M.; Furuya, T.; Kokubo, T.; Miyazaki, T.; Nakamura, T. Preparation and assessment of revised simulated body fluids. *J. Biomed. Mater. Res. Part A* **2003**, *65*, 188–195. [CrossRef]

31. Zeng, R.C.; Qi, W.C.; Cui, H.Z.; Zhang, F.; Li, S.Q.; Han, E.H. In vitro corrosion of as-extruded Mg–Ca alloys—The influence of Ca concentration. *Corros. Sci.* **2015**, *96*, 23–31. [CrossRef]
32. Wang, C.Q.; Sun, M.; Zheng, F.Y.; Peng, L.M.; Ding, W.J. Improvement in grain refinement efficiency of Mg–Zr master alloy for magnesium alloy by friction stir processing. *J. Magnes. Alloys* **2014**, *2*, 239–244. [CrossRef]
33. Zeng, Z.R.; Bian, M.Z.; Xu, S.W.; Davies, C.H.J.; Birbilis, N.; Nie, J.F. Effects of dilute additions of Zn and Ca on ductility of magnesium alloy sheet. *Mater. Sci. Eng. A* **2016**, *674*, 459–471. [CrossRef]
34. Gu, X.N.; Li, S.S.; Li, X.M.; Fan, Y.B. Magnesium based degradable biomaterials: A review. *Front. Mater. Sci.* **2014**, *8*, 200–218. [CrossRef]
35. Yang, G.L.; He, F.M.; Yang, X.F.; Wang, X.X.; Zhao, S.F. Bone responses to titanium implants surface-roughened by sandblasted and double etched treatments in a rabbit model. *Oral Surg. Oral Med. Oral Pathol. Oral Radiol. Endodontol.* **2008**, *106*, 516–524. [CrossRef] [PubMed]
36. Hagihara, K.; Fujii, K.; Matsugaki, A.; Nakano, T. Possibility of Mg-and Ca-based intermetallic compounds as new biodegradable implant materials. *Mater. Sci. Eng. C* **2013**, *33*, 4101–4111. [CrossRef] [PubMed]
37. Von Der Hoh, N.; Bormann, D.; Lucas, A.; Denkena, B.; Hackenbroich, C.; Meyer-Lindenberg, A. Influence of different surface machining treatments of magnesium-based resorbable implants on the degradation behavior in rabbits. *Adv. Eng. Mater.* **2009**, *11*, 47–54. [CrossRef]
38. Song, Y.W.; Shan, D.Y.; Han, E.H. Electrodeposition of hydroxyapatite coating on AZ91D magnesium alloy for biomaterial application. *Mater. Lett.* **2008**, *62*, 3276–3279. [CrossRef]
39. Zhang, C.Y.; Gao, J.C.; Zeng, R.C.; Liu, C.L.; Wu, X.; Wu, D. Corrosion behavior of magnesium alloy AZ31 with calcium phosphate coating in hank's solition. *J. Chin. Ceram. Soc.* **2010**, *38*, 885–891.

metals

MDPI

Article

The Distribution and Severity of Corrosion Damage at Eight Distinct Zones of Metallic Femoral Stem Implants

Roohollah Milimonfared [1], Reza H. Oskouei [1], Mark Taylor [1] and Lucian B. Solomon [2,3,*]

[1] The Medical Device Research Institute, College of Science and Engineering, Flinders University, Clovelly Park, SA 5042, Australia; sina.milimonfared@flinders.edu.au (R.M.); reza.oskouei@flinders.edu.au (R.H.O.); mark.taylor@flinders.edu.au (M.T.)

[2] Department of Orthopaedics and Trauma, Royal Adelaide Hospital, Adelaide, SA 5000, Australia

[3] Centre for Orthopaedic & Trauma Research, University of Adelaide, Adelaide, SA 5000, Australia

* Correspondence: bogdan.solomon@sa.gov.au; Tel.: +61-8-70741997

Received: 3 October 2018; Accepted: 17 October 2018; Published: 18 October 2018

Abstract: Metallic taper junctions of modular total hip replacement implants are analysed for corrosion damage using visual scoring based on different granularity levels that span from analysing the taper holistically to dividing the taper into several distinct zones. This study aims to objectively explore the spatial distribution and the severity of corrosion damage onto the surface of metallic stem tapers. An ordinal logistic regression model was developed to find the odds of receiving a higher score at eight distinct zones of 137 retrieved stem tapers. A method to find the order of damage severity across the eight zones is introduced based on an overall test of statistical significance. The findings show that corrosion at the stem tapers occurred more commonly in the distal region in comparison with the proximal region. Also, the medial distal zone was found to possess the most severe corrosion damage among all the studied eight zones.

Keywords: corrosion; metallic implants; taper junction; regression; total hip arthroplasty

1. Introduction

Despite the clinical benefits of modularity in total hip replacement (THR) implants, modular interfaces such as head-neck taper junction sustain mechanically assisted crevice corrosion due to relative micro-motions at the metallic interface and also the presence of corrosive body fluid [1,2]. Previous studies [3–5] have reported that the solid and soluble wear debris and corrosion products released from the head-neck junction may elicit untoward host body reactions such as osteolysis, peri-prosthetic fracture, and metallosis. Depending on the intensity of these postoperative complications, revision surgeries may be needed to replace failed prostheses.

Through large-scale retrieval studies, the surface damage sustained by retrieved implants is assessed, and possible associations between several implant/patients factors and the extent/location of the damage are investigated. The severity of the damage is quantified by using visual scoring methods [6,7]. To date, many studies have applied these methods (with or without modifications) to various modular junctions [7–10]. Upon scoring the damage, each study employs a causal-explanatory statistical modelling to investigate the effect of a particular set of factors (predictors) on the damage score.

In the head-neck junction, stems have a tapered geometry which can be divided into several zones (e.g., anterior, medial, posterior, and lateral quadrants). A deeper level of score granularity can provide more details about the severity and spatial distribution of damage. Distribution of the corrosion damage over the distinct zones of tapers has been investigated by a limited number of studies [8,11–19].

The number of zones scored at stem tapers has seldom gone beyond four (anterior, medial, posterior, and lateral quadrants). One reason for that could be the complexity of conducting pairwise comparisons within the groups of zone factor. With four zones, six combinations (order disregarded) would be required. If it is desired to consider the distal and proximal regions of each quadrant as well, 28 (i.e., $\frac{8!}{(8-2)!\times 2!}$) pairwise comparisons would be required to investigate the damage thoroughly. The studies that scored the distal and proximal regions separately have observed different damage patterns within these regions [15,18,19]. Therefore, it is necessary to look at stem taper zones with a higher level of granularity in order to explore whether any significant difference exists between the distal and proximal regions of the quadrants.

This study introduces a method for addressing this gap. Using this approach, eight individual corrosion scores are assigned to eight distinct zones of each metallic stem taper. Next, an ordinal logistic regression (OLR) model is used to quantitatively compare the severity of corrosion damage at these eight zones.

2. Materials and Methods

2.1. Retrieved Implants Information

This study was approved by the Southern Adelaide Clinical Human Research Ethics Committee (Reference No. 485.13); 137 total hip replacement implants retrieved between 1995 and 2015 at the Royal Adelaide Hospital (RAH), Adelaide, Australia were selected. The selection was limited to include only detached head-neck junctions so that the stem tapers were accessible for assessment. The retrieved implants had been disinfected by immersing in 70% ethanol for four days followed by a 4% Biogram solution (polyphenolic disinfectant and detergent with 18% phenol) for 48–72 h. Biologic debris (blood or proteinaceous films) had been removed using a cotton bud without abrasion. The stem tapers, selected for this research, were further cleaned with acetone followed by a gentle wipe with a soft nylon brush. Eleven implant/patient factors were retrieved from Our Patient Management and Outcomes Database (OPMOD) of the RAH. Table 1 provides the demography of these categorical and continuous factors. The missing information associated with each factor supplements the quantity of each factor to add up to 137. This study only looks at the distribution and severity of corrosion. Therefore, the missing patient and implant information did not pose any concern.

Table 1. Demographics of the selected retrievals for this study.

Predictor	Quantity (% Frequency)	Median	Range
Head Material			
CoCr	60 (43.8)		
Stainless Steel (SS)	7 (5.1)		
Ceramic	8 (5.8)		
Stem Material			
CoCr	54 (39.4)		
Stainless Steel (SS)	41 (29.9)		
Titanium	31 (22.6)		
Stem Fixation			
Cemented	76 (55.5)		
Cementless	50 (36.5)		
Gender			
Female	57 (45.2)		
Male	69 (54.8)		
Stem Taper			
12/14	52(38.0)		
V40	19 (13.9)		
9/10	12 (8.8)		

Table 1. *Cont.*

Predictor	Quantity (% Frequency)	Median	Range
6°	8 (5.8)		
C-TAPER	8 (5.8)		
TYPE 1	2 (1.5)		
11/13	3 (2.2)		
10/12	1 (0.7)		
Joint Side			
Right	69 (54.8)		
Left	57 (45.2)		
Head Diameter (mm)		28	22–55
Time to Revision (year)		6	0–35
Weight (kg)		77	51–178
Age at Primary (year)		63.5	22–85

2.2. *Visual Assessment of Corrosion Damage*

The Goldberg's scoring method [7] was used to inspect and rate corrosion on the stem tapers (Table 2). Based on this method, eight distinct zones of the retrieved stem tapers were scored individually. Fretting wear was not scored because it has been reported by several studies that fretting may be masked by corrosion damage; and therefore, hard to visually identify [14,15,19,20].

Table 2. Visual criteria for scoring corrion damage.

Score	Visual Criteria
1 (None)	No Visible Corrosion
2 (Mild)	<30% Surface Discoloured/Dull
3 (Moderate)	>30% Surface Discoloured/Dull or <10% Containing Black Debris, Pits or Etch Marks
4 (Severe)	>10% of Surface Containing Black Debris, Pits, or Etch Marks

Also, it is thought that the severity of fretting in Goldberg's method cannot be measured consistently because the pitch of the machined threads over the taper surface varies among different stem designs [14,21]. Lastly, fretting scars can be mixed up with scratches caused by attaching or detaching the head intraoperatively [7,12,21].

In order to have a consistent scoring, one trained investigator (RM) evaluated the damage. The stem tapers were visually scored twice in a random order. Each stem taper was photographed and eight zones (posterior-distal (PD), posterior-proximal (PP), medial-distal (MD), medial-proximal (MP), anterior-distal (AD), anterior-proximal (AP), lateral-distal (LD), and lateral-proximal (LP)) were identified according to our previous study [22]. Figure 1 displays an exemplary taper for each score level.

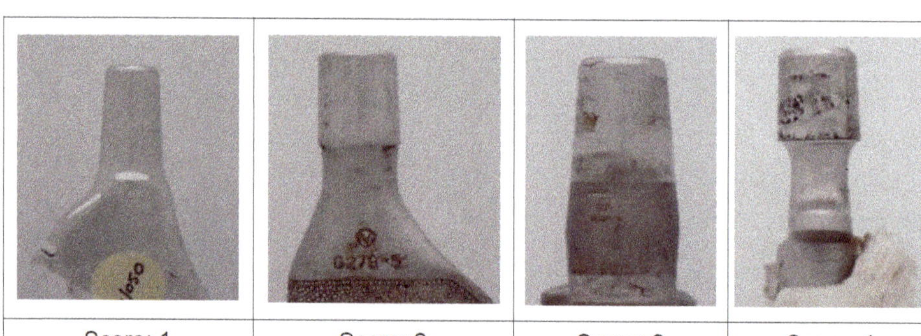

Figure 1. Corrosion damage scores of 1 through 4 for stem tapers.

2.3. Statistical Analysis

In this study, SPSS (version 25) was used for the statistical analysis and a p-value of <0.05 was determined as the level of statistical significance. Weighted kappa (κ_W) with quadratic weights was run to determine the single-observer repeatability of the corrosion scores. A confusion matrix was established to quantify the disagreements. For quadratic weights, the further away a disagreement was from the perfect agreement, the more harshly that disagreement is considered. The strength of agreement based on the magnitude of the weighted kappa (κ_W) was interpreted according to the guideline reported in Landis et al. [23].

Having an ordinal dependent variable (DV) as the response, OLR was employed to capture the ordered nature of the DV levels. The OLR model in this study uses cumulative logits. Selection of cumulative logits against other models (e.g., adjacent or continuation categories) was due to the interest of this study to use the entire response scale regardless of the score level.

Consequently, the cumulative odds OLR came with proportional odds constraint to ensure the regression lines across the DV levels are parallel. This OLR model divides the categories of the ordinal DV to run cumulative logits, as demonstrated in Table 3.

Table 3. An ordinal dependent variable (DV) with four levels giving three cumulative probabilities and consequently logits.

Binomial Regression	Event Category	Non-Event Categories
1	Probability (score ≤ 1) "none"	Probability (score > 1) "mild", "moderate", and "severe"
2	Probability (score ≤ 2) "none" and "mild"	Probability (score > 2) "moderate" and "severe"
3	Probability (score ≤ 3) "none", "mild", and "moderate"	Probability (score > 3) "severe"

With a four-level DV, this OLR model outputs three binomial logistic regressions, according to Equations (1)–(3) that predict the probability of being classified into the 'lower' categories as opposed to the 'higher' categories for each dichotomization of the ordinal DV based on the cumulative probabilities.

$$logit(success) = \ln\left(\frac{Prob(score \leq 1)}{Prob(score > 1)}\right) \quad (1)$$

$$logit(success) = \ln\left(\frac{Prob(score \leq 2)}{Prob(score > 2)}\right) \quad (2)$$

$$logit(success) = \ln\left(\frac{Prob(score \leq 3)}{Prob(score > 3)}\right) \quad (3)$$

2.3.1. The Ordinal Logistic Regression (OLR) Assumptions

Before deploying an OLR model, four assumptions (constraints) needed to be considered to ensure the validity of the results. The first assumption mandates the DV (visual scores) having an ordinal level of measurement which is valid here. Under the second assumption, there should be at least one independent variable (IV) that is continuous, ordinal or categorical (including dichotomous variables) which is valid as well.

The other two assumptions are related to the characteristics of the data. The third assumption mandates no multi-collinearity between the IVs. It was implemented by incorporating collinearity diagnostic under linear regression which returns the variance of inflation factor (VIF). VIF indicates to what extent a particular IV contributes to multi-collinearity issues within the dataset. In this study, VIF values beyond 10 were considered as having multi-collinearity as a rule of thumb.

The fourth assumption checks for having proportional odds. Here, the test of parallel lines was used to compare the fit of the proportional odds model to a model with varying slope coefficients. It was desired not to reject the null hypothesis that states the slope coefficients are the same across the three cumulative regression models. If true, the effect of each IV will be identical at each cumulative logit which is desired here.

2.3.2. Overall Parameter Estimates

As pointed out earlier, the type of OLR model used in this study produces an equation for each cumulative logit. As there are four categories of the DV, three cumulative logits (Equations (1)–(3)) are expected. Also, the assumption of proportional odds constrains the slope coefficients to be the same for all the three equations, so it is just going to be the thresholds that may vary between the three equations.

Since changes in log odds do not have much intuitive meaning, the ratio of the odds between any two categories or a unit change in a numerical IV is reported. The odds ratio (OR) was calculated as the exponential of the log odds of the slope coefficient. Also, the 95% confidence intervals of the OR and the significance levels are reported.

Unlike the numerical and dichotomous IVs, zone, as a polytomous IV, demands additional calculations to complete an overall test of statistical significance. To exhaust the entire pairwise comparison of the categories, one category was taken as the reference, and the rest were compared with that as primary categories. In each significance test, each zone had to be recoded into a new variable with the desirable reference category being coded as the last category (highest level).

3. Results

For the assessment of intra-observer repeatability, the weighted kappa (κW) with quadratic weights indicated a statistically significant agreement, κW = 0.64 (95% CI, 0.59 to 0.69), $p < 0.001$ between the two sets of scores. According to [23], the strength of the agreement was classified as good. Before using the OLR model, preliminary data analysis was carried out by looking at frequency histograms of the scores in different zones. Using various bin sizes and definitions, a number of different histograms were generated to graphically summaries the distribution of scores across the eight taper zones.

3.1. Distribution of Corrosion Scores

Visual scoring of the 137 stem tapers across the eight zones resulted in 1096 corrosion scores. Table 4 summarizes the frequency of each score level. Score level 2 had the highest quantity (512) while the lowest quantity (51) belonged to score level 4.

Table 4. The quantity of the zones having each score level.

Score	Quantity (%)
1	359 (32.8)
2	512 (46.7)
3	174 (15.9)
4	51 (4.7)

Figure 2 illustrates the distribution of the corrosion scores at each zone. This figure can be used to compare the variability of each score level across the eight zones. The values are the percentage of each score out of 137 in every zone (the percentages in each zone add up to 100%). Score levels 1 through 4 stood in the first place at zones PP, AD, MD, and MP, respectively.

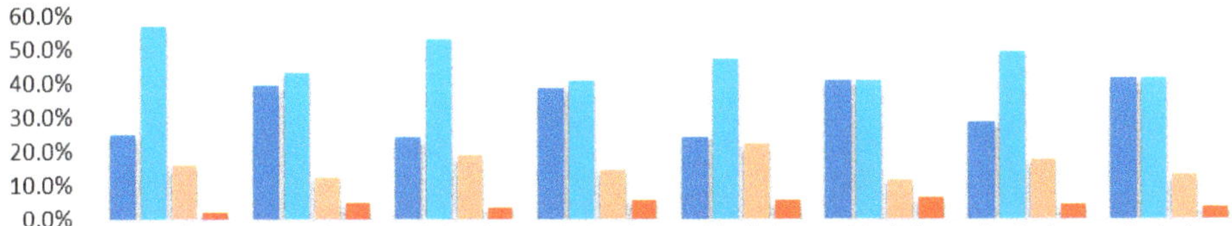

	Anterior Distal	Anterior Proximal	Lateral Distal	Lateral Proximal	Medial Distal	Medial Proximal	Posterior Distal	Posterior Proximal
1	24.8%	39.4%	24.1%	38.7%	24.1%	40.9%	28.5%	41.6%
2	56.9%	43.1%	53.3%	40.9%	47.4%	40.9%	49.6%	41.6%
3	16.1%	12.4%	19.0%	14.6%	22.6%	11.7%	17.5%	13.1%
4	2.2%	5.1%	3.6%	5.8%	5.8%	6.6%	4.4%	3.6%

Figure 2. Distribution of corrosion score levels across the entire eight stem taper zones of the 137 retrieved implants.

Considering the unbalanced score levels, the first two score levels that are higher in quantity (i.e., 359 and 512) always show higher percentages compared with score levels 3 and 4 within each zone.

To better compare the severity of damage across the zones, two more configurations of scores (by combining the original score levels) were also explored. The first configuration groups the first and the last two score levels into low and high groups, respectively. Figure 3 visualizes this configuration and compares each score group across the eight zones.

	Anterior Distal	Anterior Proximal	Lateral Distal	Lateral Proximal	Medial Distal	Medial Proximal	Posterior Distal	Posterior Proximal
1-2	81.8%	82.5%	77.4%	79.6%	71.5%	81.8%	78.1%	83.2%
3-4	18.2%	17.5%	22.6%	20.4%	28.5%	18.2%	21.9%	16.8%

Figure 3. The quantity of the double score levels at each zone (scores 1 and 2 versus scores 3 and 4 combined).

As expected, the low score group which comprises (359 + 512) scores has a higher frequency compared with the high score group (174 + 51). This configuration can better show which zones have more severe corrosion damage (for example, MD and LD zones). Also, at zones MD and PP, the smallest and largest gaps between these two combined score levels were observed.

The third configuration preserves score level 1 and combines the other three score levels to form two new score groups of intact and corroded stem tapers. Figure 4 illustrates the frequencies of these two score groups.

The medial distal zone had the largest difference between these two score groups which confirms that this particular zone is most damaged. Also, the posterior-proximal zone had the smallest difference between the two score groups (thus least damaged). As a key finding, the distal regions of the four quadrants showed more corrosion damage compared with the proximal regions.

These finding from the histogram can shed light on the likely outcome of the OLR model. In particular, when the number of DV levels are higher, cumulative logits models may become infeasible. Histograms can determine which score levels are more important to be compared via using other types of OLR models such as adjacent categories.

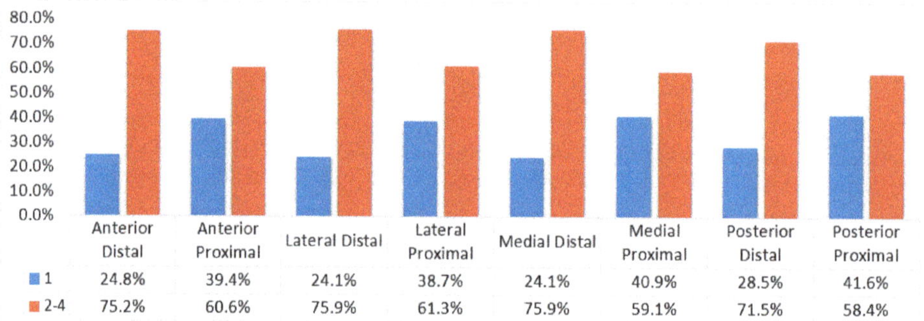

Figure 4. Distribution of corroded stem tapers against the intact group.

3.2. Comparison of Corrosion in the Zones

Cumulative odds OLR with proportional odds was employed to conduct pairwise comparisons between the zones. First, it was established whether zone is statistically significant overall. From the test of model performed on SPSS, zone was observed to be a statistically significant ($p = 0.002$) predictor of corrosion scores in this univariate regression model.

Since no specific zone was preferential to investigate, 28 pairwise comparisons had to be undertaken which incurred additional calculations to obtain the overall omnibus statistical test. Table 5 summarizes the OR, p-values, and confidence intervals. Significant OR values are highlighted in grey. In this table, each zone has been used seven times either as the primary or reference (inside brackets) group to exhaust the combinations. OR values below 1 indicate that for the primary category, the odds of having a higher corrosion score is lower than that of the reference category.

Table 5. The odds of observing a higher corrosion score at a primary zone compared with a reference zone.

Zone Pair	OR	p-Value	CI ($p < 0.05$)	
AD (AP)	1.493	0.077	0.957	2.329
AD (LD)	0.882	0.577	0.566	1.372
AD (LP)	1.365	0.169	0.876	2.128
AD (MD)	0.755	0.212	0.485	1.175
AD (MP)	1.524	0.063	0.977	2.378
AD (PD)	0.998	0.993	0.641	1.554
AD (PP)	1.634	0.031	1.047	2.551
AP (LD)	0.590	0.020	0.379	0.921
AP (LP)	0.914	0.693	0.586	1.427
AP (MD)	0.505	0.003	0.324	0.789
AP (MP)	1.021	0.928	0.654	1.594
AP (PD)	0.668	0.076	0.429	1.043
AP (PP)	1.094	0.692	0.701	1.709
LD (LP)	1.549	0.054	0.993	2.414
LD (MD)	0.856	0.490	0.550	1.331
LD (MP)	1.729	0.016	1.108	2.697
LD (PD)	1.132	0.583	0.727	1.762
LD (PP)	1.853	0.007	1.187	2.894
LP (MD)	0.553	0.009	0.355	0.862

Table 5. *Cont.*

Zone Pair	OR	p-Value	CI ($p < 0.05$)	
LP (MP)	1.116	0.628	0.715	1.742
LP (PD)	0.731	0.167	0.469	1.140
LP (PP)	1.197	0.429	0.767	1.869
MD (MP)	2.019	0.002	1.294	3.152
MD (PD)	1.322	0.216	0.850	2.058
MD (PP)	2.165	0.001	1.386	3.382
MP (PD)	0.655	0.062	0.420	1.022
MP (PP)	1.072	0.760	0.686	1.675
PD (PP)	1.637	0.030	1.049	2.556

The reciprocal of odds ratios can be calculated to compare a reference group with a primary group. To compare the severity of corrosion across the entire eight zones, the odds ratios were sorted and plotted (Figure 5). The red and blue bars indicate the significant and insignificant OR values, respectively. An OR equal to 1 indicates equal odds of observing a higher corrosion score at the primary and reference zone groups. By moving away from unity, the odds ratios that are first insignificant later on become significant. The speed by which this transition takes place is a function of the presumed statistical significance level.

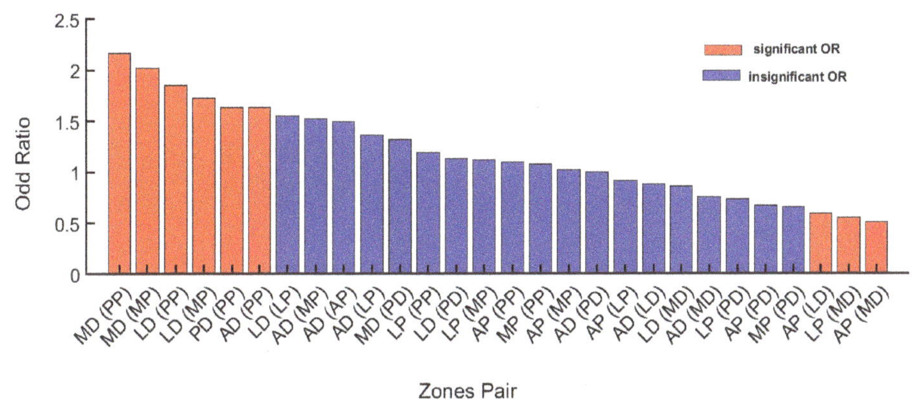

Figure 5. The 28 odds ratios sorted and colour-coded for 28 pairwise comparisons.

The severity of corrosion at each zone with respect to the other zones was assessed based on its corresponding OR values. For each zone, Table 5 has provided seven OR values wherein that particular zone appears as either primary or reference.

Table 6 sorts the eight zones from the least to the most severely damaged according to the value of C1 + C2. This value quantifies how many times each zone had a higher likelihood of damage compared with the other seven zones throughout the 28 pairwise comparisons. C1 indicates how many time a particular zone, as the primary, had an OR value above 1, while C2 indicates how many times that same zone, as the reference, had an OR value below 1. Therefore, both C1 and C2 reflects the frequency of each zone appearing as more severely damaged with respect to the other zones.

Table 6. The frequency of each zone showing statistically significant odds ratio (OR).

Zone	C1	C2	C1 + C2
Posterior Proximal (PP)	0	0	0
Medial Proximal (MP)	1	0	1
Anteriori Proximal (AP)	2	0	2

Table 6. *Cont.*

Zone	C1	C2	C1 + C2
Lateral Proximal (LP)	2	1	3
Anterior Distal (AD)	4	0	4
Posterior Distal (PD)	1	4	5
Lateral Distal (LD)	4	2	6
Medial Distal (MD)	3	4	7

Zones PP and MD were identified having the least and highest severity of corrosion. Interestingly, proximal and distal regions were found to be grouping together in this table with the distal region showing more damage compared with the proximal region across the four quadrants in the studied stem tapers.

4. Discussion

Eight distinct zones of the stem tapers including anterior-distal, anterior-proximal, medial-distal, medial-proximal, posterior-distal, posterior-proximal, lateral-distal, and lateral-proximal were scored and statistically compared to identify the zone(s) with the most severe corrosion damage in the retrieved implants studied in this work. It is noted that there are several studies in the literature that chose to score stem tapers holistically, not locally [9,10,24–26].

Within the studies [11,12,15–19] that scored stem tapers locally, the pools of implants had a limited diversity in terms of implant properties (e.g., head diameter, articulation type, and stem design). Therefore, it was deemed necessary to explore whether a similar distribution of corrosion damage can be seen in a more heterogeneous pool of implants.

To the best of our knowledge, there are only two studies [18,19] in the literature that, similar to this work, have assigned eight local scores to the stems with the rest using lower numbers of zones. In those two studies, one did not compare the scores between the zones [18]. The other compared the four quadrants, and the two distal and proximal regions separately in terms of corrosion severity and did not determine which zone(s) had the most severe damage [19].

Routine causal-explanatory statistical analyses require only one score as the descriptor of damage for each implant. The majority of these studies have chosen to combine the local scores by calculating an overall value [8,11–14]. This approach has led to the presumption that this global score is a continuous variable; and, thus, the statistical analyses for continuous variables have been utilised. Analysing a continuous variable with an interval or ratio level of measurement is generally less complex in nature. However, an increased number of levels in the global score does not necessarily imply a known "distance" between the score levels. Therefore, this approach was treated with suspicion in this study and was not adopted.

Here, the corrosion scores were analysed using a univariate OLR model, and the odds ratios along with their p-values were reported. Since there was no particular hypothesis about the relative level of corrosion at the eight zones, 28 pairwise comparisons were carried out to exhaust the entire pairwise comparison of the zones. The distal region of the medial quadrant was found to have the highest odds of receiving a higher corrosion score which is aligned with the previous findings in the literature that identified the distal region [19,20,27] and the medial quadrant [7,10,16,28] having the highest corrosion scores. Also, this study shows that the distal region of all the four quadrants had more corrosion damage in comparison with the proximal region of those quadrants. Therefore, it was found that, regardless of the quadrant, corrosion damage is more present distally than proximally.

Generally, the higher severity of wear or corrosion at a specific zone has been attributed to several factors such as increased micro-motions at the interface, head or stem materials, head diameter, high friction moments, and poor lubrication of the bearing articulation. While some act as root causes, the others play the role of causal factors. Also, damage at the head-neck taper junction usually appears as a combination of wear and corrosion mechanisms. Some of these factors may only contribute to a specific mode of damage, while others may contribute towards a set of damage mechanisms.

In a retrieval study of 231 implants [7] the stem tapers received four fretting and corrosion scores corresponding to the four quadrants. The medial and lateral scores were observed to be significantly higher than the scores at the other two quadrants (posterior and anterior). This was explained to be due to a higher likelihood of micro-motions between the head and neck about an axis in the sagittal plane. Similar to the present study, the pool of implants in this work had a wide diversity, and higher corrosion scores at the medial quadrant suggest that it could be a phenomenon independent of the included patient and implant factors.

Wilson et al. [29] explained how at the double-tapered cone design of Profemur Z, the proximal end of the neck experiences an almost pure compression and shear loading. High frictional moments at taper junctions were related to poor lubrication of the articulation interfaces by another study [30].

The medial quadrant was identified to have higher corrosion scores in a retrieval study of 52 S-ROM components [16]. It was hypothesised that greater micro-motions at this quadrant could result in a more frequent disruption of the passive oxide layer; and consequently, more severe corrosion damage. Similar to the conclusion of the Wilson et al. [29] study, they reported that this region is generally under a compression-loading regime. A computational modelling of the stem taper stresses paired with large diameter heads confirmed this hypothesis after witnessing maximum levels of principal stresses at the medial quadrant [31]. In that work, a 3D model of a 12/14 titanium taper was paired with cobalt-chromium and alumina heads. Increasing the head diameter increased this quadrant's stresses distal to the junction significantly. It was highlighted that the pairing of a small taper and a large head leads to a larger moment arm transmitting a higher force to a small surface area which facilitates tribo-corrosion.

A relatively higher amount of load and stress at the medial quadrant causes elastic strains which appear as surface compression. This condition may lead to micro-motions of approximately 5 to 40 μm [32] which in turn may result in abrasion or fracture of the oxide layer. The subsequent changes in the metal surface potential and the continuous re-passivation of the oxide layer change the chemistry of the crevice solution. Ultimately, the deaeration and pH decrease of the solution initiate crevice corrosive attacks [33,34]. Crevice corrosion has been reported to occur near the bore opening which may explain observing more severe corrosion at the distal region [35].

Besides micro-motions, galvanic corrosion at this interface due to using mixed metal components is a potential source of material loss. In this study, 18 (13.1%) implants had mixed head and stem materials, whereas 45 (32.8%) had similar materials. Therefore, galvanic corrosion cannot be nominated as the sole mechanism of corrosion.

These studies have used relatively homogenous pools of implants, yet they observed higher levels of corrosion at the medial quadrant or distal zones of stem tapers. Based on the findings of the present study which shows that the distal region of the medial quadrant sustains the most severe corrosion damage, it is understood that this particular zone is most severely damaged versus all the other zones regardless of the properties and patient characteristics of the investigated pool of implants.

5. Conclusions

This study introduces a method to statistically compare the severity of corrosion damage at eight distinct zones of stem tapers. A pool of 137 retrieved total hip replacement implants was visually scored at the head-neck junction for corrosion damage using Goldberg's method. An OLR model with proportional odds was used to determine the odds of observing a higher score at a primary zone compared with a reference zone. The findings of this study can be highlighted as below:

- The corrosion score level 2 was observed having the highest frequency (46.7%).
- Posterior-proximal and medial-distal were identified as the zones with the least and most severity of corrosion.

- Interestingly, the proximal and distal regions were found to be grouping together with the distal region showing more damage in comparison with the proximal region of the four quadrants.
- Out of the 28 pairwise comparisons of these eight zones, nine pairs of zones were identified to be significantly different regarding corrosion damage. This observation objectively shows the high diversity in corrosion damage across these zones.
- Retrieval studies of taper junctions are, therefore, recommended to score the zones separately and avoid adding up local scores to be used with an interval or ratio level of measurement.

Author Contributions: R.M. studied the implants, conducted the statistical analysis, discussed the results and drafted the manuscript. R.H.O., M.T. and L.B.S. supervised and contributed to all the project tasks including revising and writing of the paper.

Funding: This research received no external funding.

Acknowledgments: The authors would like to acknowledge the support through an Australian Government Research Training Program Scholarship for Roohollah Milimonfared's PhD. The authors also thank the Department of Orthopaedics and Trauma, Royal Adelaide Hospital (Adelaide, Australia), and in particular, Kerry Costi and Roumen Stamenkov for their assistance in accessing the retrieved implants.

Conflicts of Interest: The authors have no conflicts of interest to declare.

References

1. Hussenbocus, S.; Kosuge, D.; Solomon, L.B.; Howie, D.W.; Oskouei, R.H. Head-neck taper corrosion in hip arthroplasty. *BioMed Res. Int.* **2015**, *2015*, 758123. [CrossRef] [PubMed]
2. Oskouei, R.H.; Barati, M.R.; Farhoudi, H.; Lucian, M.T.; Solomon, B. A new finding on the in-vivo crevice corrosion damage in a CoCrMo hip implant. *Mat. Sci. Eng. C* **2017**, *79*, 390–398. [CrossRef] [PubMed]
3. Jacobs, J.J.; Cooper, H.J.; Urban, R.M.; Wixson, R.L.; Della Valle, C.J. What do we know about taper corrosion in total hip arthroplasty? *J. Arthroplasty* **2014**, *29*, 668–669. [CrossRef] [PubMed]
4. Cooper, H.J.; Della Valle, C.J.; Berger, R.A.; Tetreault, M.; Paprosky, W.G.; Sporer, S.M.; Jacobs, J.J. Corrosion at the head-neck taper as a cause for adverse local tissue reactions after total hip arthroplasty. *J. Bone Jt. Surg. Am.* **2012**, *94*, 1655–1661. [CrossRef]
5. Vundelinckx, B.J.; Verhelst, L.A.; De Schepper, J. Taper corrosion in modular hip prostheses: Analysis of serum metal ions in 19 patients. *J. Arthroplasty* **2013**, *28*, 1218–1223. [CrossRef] [PubMed]
6. Gilbert, J.L.; Buckley, C.A.; Jacobs, J.J. In vivo corrosion of modular hip prosthesis components in mixed and similar metal combinations. The effect of crevice, stress, motion, and alloy coupling. *J. Biomed. Mater. Res.* **1993**, *27*, 1533–1544. [CrossRef] [PubMed]
7. Goldberg, J.R.; Gilbert, J.L.; Jacobs, J.J.; Bauer, T.W.; Paprosky, W.; Leurgans, S. A multicenter retrieval study of the taper interfaces of modular hip prostheses. *Clin. Orthop. Relat. Res.* **2002**, *401*, 149–161. [CrossRef]
8. Hothi, H.S.; Berber, R.; Panagiotopoulos, A.C.; Whittaker, R.K.; Rhead, C.; Skinner, J.A.; Hart, A.J. Clinical significance of corrosion of cemented femoral stems in metal-on-metal hips: A retrieval study. *Int. Orthop.* **2016**, *40*, 2247–2254. [CrossRef] [PubMed]
9. Carlson, J.C.H. Femoral stem fracture and in vivo corrosion of retrieved modular femoral hips. *J. Arthroplasty* **2012**, *27*, 1389–1396. [CrossRef] [PubMed]
10. Higgs, G.B. Method of characterizing fretting and corrosion at the various taper connections of retrieved modular components from metal-on-metal total hip arthroplasty. Metal-on-metal total hip replacement devices. *ASTM Int.* **2013**, *1560*, 146–156. [CrossRef]
11. Tan, S.C.; Lau, A.C.; Del Balso, C.; Howard, J.L.; Lanting, B.A.; Teeter, M.G. Tribocorrosion: Ceramic and oxidized zirconium vs cobalt-chromium heads in total hip arthroplasty. *J. Arthroplasty* **2016**, *31*, 2064–2071. [CrossRef] [PubMed]
12. Tan, S.C.; Teeter, M.G.; Del Balso, C.; Howard, J.L.; Lanting, B.A. Effect of taper design on trunnionosis in metal on polyethylene total hip arthroplasty. *J. Arthroplasty* **2015**, *30*, 1269–1272. [CrossRef] [PubMed]
13. Hothi, H.S.; Berber, R.; Whittaker, R.K.; Bills, P.J.; Skinner, J.A.; Hart, A.J. Detailed inspection of metal implants. *Hip Int.* **2015**, *25*, 227–231. [CrossRef] [PubMed]

14. Hothi, H.S.; Matthies, A.K.; Berber, R.; Whittaker, R.K.; Skinner, J.A.; Hart, A.J. The reliability of a scoring system for corrosion and fretting, and its relationship to material loss of tapered, modular junctions of retrieved hip implants. *J. Arthroplasty* **2014**, *29*, 1313–1317. [CrossRef] [PubMed]
15. De Martino, I.; Assini, J.B.; Elpers, M.E.; Wright, T.M.; Westrich, G.H. Corrosion and fretting of a modular hip system: A retrieval analysis of 60 rejuvenate stems. *J. Arthroplasty* **2015**, *30*, 1470–1475. [CrossRef] [PubMed]
16. Munir, S.; Michael, B.C.; Christina, E.; Anna, S.; William, L.W. Corrosion in modular total hip replacements: An analysis of the head–neck and stem–sleeve taper connections. *Semin. Arthroplasty* **2013**, *24*, 240–245. [CrossRef]
17. Stamer, C.M. Assessment of Bore-Cone Taper Junctions in Explanted Modular Total Hip Replacements. Master's Thesis, Clemson University, Clemson, SC, USA, 2015.
18. Kao, Y.-Y.J.; Koch, C.N.; Wright, T.M.; Padgett, D.E. Flexural rigidity, taper angle, and contact length affect fretting of the femoral stem trunnion in total hip arthroplasty. *J. Arthroplasty* **2016**, *31*, S254–S258. [CrossRef] [PubMed]
19. Triantafyllopoulos, G.K.; Elpers, M.E.; Burket, J.C.; Esposito, C.I.; Padgett, D.E.; Wright, T.M. Otto aufranc award: Large heads do not increase damage at the head-neck taper of metal-on-polyethylene total hip arthroplasties. *Clin. Orthop. Relat. Res.* **2015**, *474*, 330–338. [CrossRef] [PubMed]
20. Gonzalez, J.L., Jr. In-Vivo Corrosion and Fretting of Modular TI-6AL-4V/CO-CR-MO Hip Prostheses: The Influence of Microstructure and Design Parameters. Master's Thesis, Florida International University, Miami, FL, USA, 2015.
21. Kocagöz, S.B.; Underwood, R.J.; Sivan, S.; Gilbert, J.L.; MacDonald, D.W.; Day, J.S.; Kurtz, S.M. Does taper angle clearance influence fretting and corrosion damage at the head–stem interface? A matched cohort retrieval study. *Semin. Arthroplasty* **2013**, *24*, 246–254. [CrossRef] [PubMed]
22. Milimonfared, R.; Oskouei, R.H.; Taylor, M.; Solomon, L.B. An intelligent system for image-based rating of corrosion severity at stem taper of retrieved hip replacement implants. *Med. Eng. Phys.* **2018**. [CrossRef] [PubMed]
23. Landis, J.R.; Koch, G.G. The measurement of observer agreement for categorical data. *Biometrics* **1977**, *33*, 159–174. [CrossRef] [PubMed]
24. Higgs, G.B.; MacDonald, D.W.; Lowell, J.; Padayatil, A.; Mihalko, W.M.; Siskey, R.L.; Gilbert, J.L.; Rimnac, C.M.; Kurtz, S.M. Is corrosion a threat to the strength of the taper connection in femoral components of total hip replacements? *Corrosion* **2017**, *73*, 1538–1543. [CrossRef]
25. Whittaker, R.K.; Hothi, H.S.; Meswania, J.M.; Berber, R.; Blunn, G.W.; Skinner, J.A.; Hart, A.J. The effect of using components from different manufacturers on the rate of wear and corrosion of the head-stem taper junction of metal-on-metal hip arthroplasties. *Bone Jt. J.* **2016**, *98*, 917–924. [CrossRef] [PubMed]
26. Matthies, A.K.; Racasan, R.; Bills, P.; Blunt, L.; Cro, S.; Panagiotidou, A.; Blunn, G.; Skinner, J.; Hart, A.J. Material loss at the taper junction of retrieved large head metal-on-metal total hip replacements. *J. Orthop. Res.* **2013**, *31*, 1677–1685. [CrossRef] [PubMed]
27. Langton, D.J.; Sidaginamale, R.P.; Joyce, T.J.; Meek, R.D.; Bowsher, J.G.; Deehan, D.; Nargol, A.V.F.; Holland, J.P. A comparison study of stem taper material loss at similar and mixed metal head-neck taper junctions. *Bone Jt. J.* **2017**, *99*, 1304–1312. [CrossRef] [PubMed]
28. Higgs, G.B.; Hanzlik, J.A.; MacDonald, D.W.; Gilbert, J.L.; Rimnac, C.M.; Kurtz, S.M. Is increased modularity associated with increased fretting and corrosion damage in metal-on-metal total hip arthroplasty devices?: A retrieval study. *J. Arthroplasty* **2013**, *28*, 2–6. [CrossRef] [PubMed]
29. Wilson, D.A.; Dunbar, M.J.; Amirault, J.D.; Farhat, Z. Early failure of a modular femoral neck total hip arthroplasty component a case report. *J. Bone Jt. Surg. Am.* **2010**, *92*, 1514–1517. [CrossRef] [PubMed]
30. Bishop, N.E.; Hothan, A.; Morlock, M.M. High friction moments in large hard-on-hard hip replacement bearings in conditions of poor lubrication. *J. Orthop. Res.* **2013**, *31*, 807–813. [CrossRef] [PubMed]
31. Lavernia, C.J.; Iacobelli, D.A.; Villa, J.M.; Jones, K.; Gonzalez, J.L.; Jones, W.K. Trunnion-head stresses in THA: Are big heads trouble? *J. Arthroplasty* **2015**, *30*, 1085–1088. [CrossRef] [PubMed]
32. Porter, D.A.; Urban, R.M.; Jacobs, J.J.; Gilbert, J.L.; Rodriguez, J.A.; Cooper, H.J. Modern trunnions are more flexible: A mechanical analysis of THA taper designs. *Clin. Orthop. Relat. Res.* **2014**, *472*, 3963–3970. [CrossRef] [PubMed]
33. Jacobs, J.J.; Gilbert, J.L.; Urban, R.M. Corrosion of metal orthopaedic implants. *J. Bone Jt. Surg. Am.* **1998**, *80*, 268–282. [CrossRef]

34. Osman, K.; Panagiotidou, A.P.; Khan, M.; Blunn, G.; Haddad, F.S. Corrosion at the head-neck interface of current designs of modular femoral components essential questions and answers relating to corrosion in modular head-neck junctions. *Bone Jt. J.* **2016**, *98*, 579–584. [CrossRef] [PubMed]
35. Cook, S.D.; Barrack, R.L.; Clemow, A.J.T. Corrosion and wear at the modular interface of uncemented femoral stems. *J. Bone Jt. Surg. Br.* **1994**, *76*, 68–72. [CrossRef]

metals

MDPI

Article

The Influence of Assembly Force on the Material Loss at the Metallic Head-Neck Junction of Hip Implants Subjected to Cyclic Fretting Wear

Khosro Fallahnezhad [1], Reza H. Oskouei [1,*], Hojjat Badnava [2] and Mark Taylor [1]

[1] College of Science and Engineering, Flinders University, Clovelly Park, SA 5042, Australia; khosro.fallahnezhad@flinders.edu.au (K.F.); mark.taylor@flinders.edu.au (M.T.)

[2] Discipline of Mechanical Engineering, Behbahan University of Technology, Behbahan 6361647189, Iran; ha.badnava@me.iut.ac.ir

* Correspondence: reza.oskouei@flinders.edu.au; Tel.: +61-8-8201-2782

Received: 2 March 2019; Accepted: 3 April 2019; Published: 8 April 2019

Abstract: The impaction force required to assemble the head and stem components of hip implants is proven to play a major role in the mechanics of the taper junction. However, it is not clear if the assembly force could have an effect on fretting wear, which normally occurs at the junction. In this study, an adaptive finite element model was developed for a CoCr/CoCr head-neck junction with an angular mismatch of 0.01° in order to simulate the fretting wear process and predict the material loss under various assembly forces and over a high number of gait cycles. The junction was assembled with 2, 3, 4, and 5 kN and then subjected to 1,025,000 cycles of normal walking gait loading. The findings showed that material removal due to fretting wear increased when raising the assembly force. High assembly forces induced greater contact pressures over larger contact regions at the interface, which, in turn, resulted in more material loss and wear damage to the surface when compared to lower assembly forces. Although a high assembly force (greater than 4 kN) can further improve the initial strength and stability of the taper junction, it appears that it also increases the degree of fretting wear. Further studies are needed to investigate the assembly force in the other taper designs, angular mismatches, and material combinations.

Keywords: fretting wear; CoCrMo alloys; assembly force; material loss; modular hip implants; finite element

1. Introduction

Modular junctions are commonly used in orthopaedics, such as the head-neck taper junction in total hip replacement, to allow flexibility at the time of surgery. The aim is to generate a rigid connection between the modular components. However, fretting wear occurs due to the small-amplitude relative motions that occur at the interface under physiological loads in the body. In addition, fretting wear can remove the passive oxide layer of the metal alloy, and, with the presence of the body fluid, re-passivation occurs within the small crevice at the junction, which causes fretting corrosion [1,2]. These phenomena produce particulate debris and metal ions, which, in turn, can cause adverse local tissue reactions and, ultimately, clinical failure [3–5].

The mechanical behavior of the taper connection is dependent on a number of parameters. The material combination used at the taper [6] and the taper mismatch angle [6–9] can be controlled by the design and the manufacturing process. The type of the mechanical load as a result of daily activities can also influence the mechanical behavior of the taper junction [10,11]. Since the components are assembled intraoperatively, the assembly force is important not only to avoid loosening and dissociation after implantation [12,13] but also to establish a favorable mechanical environment to minimize fretting. There is variation between the manufacturers' recommendations on how to impact the femoral head [14], from a single light tap to several sharp hammer blows. The impaction forces generated by surgeons can vary significantly from approximately 300 N to an excess of 7500 N [15]. To date, the majority of studies [12,16–18] investigating the influence of the assembly force have focused on the dissociation force as the metric to assess the performance of the taper. Assembly forces from 2 kN to 15 kN have been investigated and a linear relationship with the dissociation force has been reported [12,16]. The dissociation force is always lower and varies between 42% [17] and 91% [18] of the assembly force. Rehmer et al. [12] reported a similar linear relationship between the assembly force and the twist off torque. Increasing the impaction force has also been shown to increase the contact area [6,19] and reduce the micro-motion [18,20] between the head and the trunnion. These studies generally suggest that a high assembly force can achieve a high degree of initial stability and fixation in the head-neck junction to more reliably withstand mechanical loads of daily activities without disconnection.

However, very few studies have attempted to address the important question of whether the assembly force has an influence on the material removal by fretting wear over an extended period. Bitter et al. [18] developed a combined experimental and finite element (FE) study to analyze the influence of assembly force (2, 4, and 15 kN) on the fretting wear of a Ti-6Al-4V femoral stem in contact with a Ti-6Al-4V taper adaptor. Their experimental results showed large standard deviations in terms of volumetric wear and no significant difference was found between the three tested assembly forces. However, it was reported that, when increasing the assembly force, the fretting wear reduces at the taper interface. They also used a simplified FE modelling approach simulating accelerated fretting (did not incorporate geometry updates to account for material loss). Employing a simplified version of the Archard equation, they defined a total wear score for the interface using the contact pressure and relative micro-motion for each contact node. As a major simplification, their model was not able to track the fretting wear process over several cycles of sliding. No correlation was found between the predicted wear scores (from the FE analysis) and the experimental volumetric wear [18]. In another FE analysis, English et al. [21] modelled a CoCr head and a titanium neck with a zero angular mismatch to estimate the material loss and contact pressure at the junction subjected to two million cycles of walking gait loading. This work was extended to explore the influence of assembly force, and they reported that higher assembly forces resulted in lower fretting wear [22]. However, they still used the critical simplification of zero angular mismatch for the junction in the dry condition. The materials modelled in the previous studies did not include the common combination of CoCr head and CoCr neck. Furthermore, the existing taper angle mismatch between the head and neck components has been ignored in the previous fretting wear studies while the angular mismatch has been found to significantly influence the mechanics of the junction [6,23]. Therefore, it could have a significant effect on fretting wear as a mechanically driven process. More importantly, the previous FE simulations have often assumed a dry condition for the contacting materials of the junction. However, the existence of body fluid at the interface of the junction may control the frictional characteristics and wear characteristics, which may influence the fretting wear behavior.

This work aims to simulate the fretting wear process and predict the material removal in a CoCr/CoCr head-neck junction through an adaptive finite element modelling approach. To achieve more realistic outcomes, the taper junction was modelled to have a distal contact with a real angular mismatch between the head and neck with the presence of simulated physiological body fluid. The main research objective was to evaluate the effect of assembly force on the material loss and fretting wear process in this type of taper junction.

2. Methods

A 3D model of a 32-mm diameter CoCr head with a 12/14 CoCr neck (as per dimensions reported by Rehmer et al. [12]) was first analyzed under normal walking load profiles [24,25] in order to determine the most critical plane of the taper junction in terms of two important fretting wear parameters of normal contact stress (contact pressure) and micro-motion. An elastic-plastic material model was used for CoCr (ISO 5832-12) with a Young's modulus of 210 GPa, Poisson's ratio of 0.30, yield stress of 910 GPa, ultimate tensile stress of 1350 GPa, and tensile elongation of 15%. The contact pressure and displacement were retrieved for all the nodes in the contacting regions of the head and neck, using a Python script, under the maximum force and moment of the loading profile. The relative displacements (micro-motions) of the contacting nodes were then determined using a MATLAB code. The middle plane passing through the superolateral region of the neck was found as the most critical plane to have the largest area of contact pressure (as shown in Figure 1) together with micro-motions. This critical plane was then employed for the main 2D fretting wear model that will be described next.

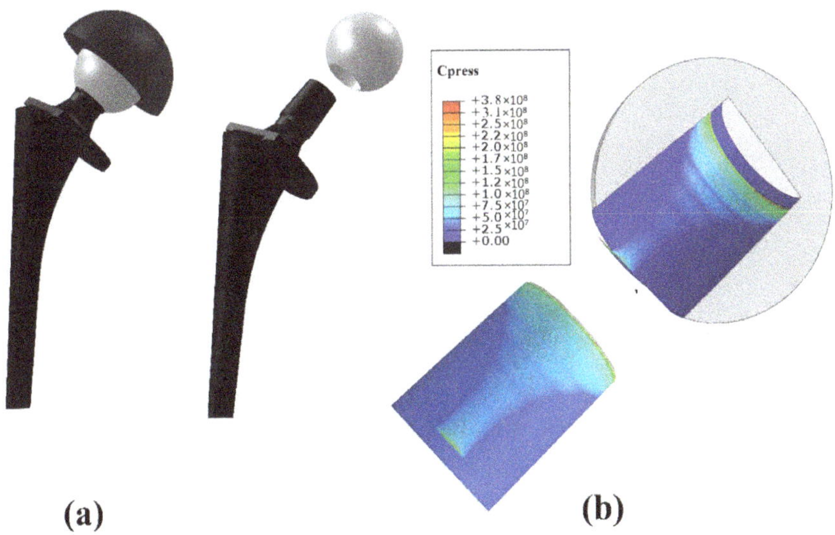

Figure 1. (**a**) A three-dimensional model of a hip joint implant and (**b**) distribution of contact pressure (in Pa) in both the head and neck under a normal walking gait loading, which indicates that the super-lateral region has the largest contacting area with both contact pressures and micro-motions.

2.1. *Fretting Wear Model Development*

The Archard wear formulation (Equation (1)) [26] was used in the fretting wear model of this work:

$$\frac{V}{S} = k\frac{F}{H} \tag{1}$$

where V (m^3) is the lost volume, S (m) is the amplitude of sliding, k (Pa^{-1}) is the wear coefficient, F (N) is the normal force, and H is the hardness of the material [26]. The Archard formulation can be localized and applied to the points of a contact region, which makes it suitable for FE simulations. In addition, the Archard equation has been previously used for fretting wear simulations with success [27–30].

Dividing both sides of the Archard wear equation (Equation (1)) by the area, yields the following.

$$h = K{\cdot}S{\cdot}p{\cdot}\Delta N \tag{2}$$

where h (m) is the depth of wear, K is the wear coefficient-to-hardness ratio (k/H), p (Pa) is the normal contact stress, and ΔN is the load cycle update interval. The main reason to re-write Equation (1) in the form of Equation (2) was to apply the Archard formula to the contacting nodes in the FE model. For the fretting wear model in the present study, a FORTRAN code was developed to trace and determine the positioning of the contacting nodes through the ABAQUS UMESHMOTION subroutine within an adaptive meshing constraint.

2.1.1. Verification

McColl et al. [29] and Ding et al. [28] developed an algorithm based on the Archard equation in order to simulate fretting wear for a pin-on-disc testing system. They reported surface profiles of the disc after various cycles of fretting wear. In order to verify the UMESHMOTION code developed for the head-neck junction in this study, a pin-on-disc model was first generated to replicate the Ding's model. This model had a very similar configuration (materials, geometry, element sizes, meshing structure, normal force, and sliding amplitude and frequency). The surface of the disc after the fretting wear process was evaluated (Figure 2) and compared with the results reported by Ding et al. [28]. Table 1 provides a comparison between the results of this study and those presented by Ding et al. [28] in terms of the width and height of the wear profile for the disc, which shows a very good level of agreement and verifies the UMESHMOTION code and its accuracy used in this study.

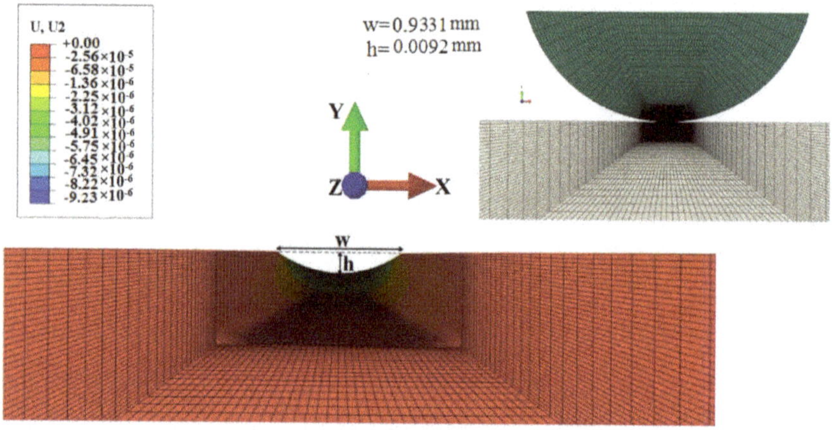

Figure 2. The distribution of displacement in the Y-axis (in meters) representing material removal from the surface of the disc under a normal force of 1200 N and a sliding amplitude of 10 μm after 18,000 cycles of fretting wear.

Table 1. The width and height of the wear profile for the disc after various fretting wear test cycles. A comparison between the computational results of this work and those presented by Ding et al. [28].

Comparison	Wear Profile Parameters on the Surface of the Disc	After 1000 Cycles	After 5000 Cycles	After 18,000 Cycles
Results of this work	Width, w (mm)	0.3512	0.7123	0.9331
	Height, h (mm)	0.0013	0.0042	0.0092
Results reported by	Width, w (mm)	0.3834	0.7644	0.9754
	Height, h (mm)	0.0013	0.0042	0.0092

2.1.2. Fretting Wear Model for the Head-Neck Taper Junction

The most critical plane of the head-neck junction that was previously identified by the 3D FE analysis was used to develop a 2D fretting wear model for the taper junction. Figure 2 illustrates the mesh structure of the 2D head-neck junction model and the profile of the corresponding force components (from the normal walking gait cycle) applied to this plane.

For the Archard wear equation (Equation (2)), the wear coefficient-to-hardness ratio (K) for the CoCr/CoCr head-neck combination was determined from a set of experimental results reported by Maruyama et al. [31]. They employed CoCr/CoCr pin-on-disc experiments under various normal contact stresses and sliding cycles in a phosphate buffered saline (PBS) condition. From their results and using the Archard equation, the wear coefficient-to-hardness ratio was determined for nine cases tested in their study. The nine K values were found to be very close with a maximum difference of 10% from which an average of $K = 1.7 \times 10^{-15}$ Pa^{-1} was calculated and used for the CoCr/CoCr taper junction model.

The coefficient of friction between CoCr and CoCr in the PBS condition was also obtained from the results reported by Maruyama et al. [31]. Their results for different contact stresses and cycles showed that the friction coefficient becomes constant at 0.60 after approximately 5000 cycles, which was used in the FE simulations of this work.

The authors' previous work [23] showed that head-neck taper junctions with distal angular mismatches have generally a better resistance to fretting wear when compared to junctions with proximal angular mismatches. Hence, in this paper, a small yet realistic distal angular mismatch (0.01°) was chosen for all the cases in order to investigate the influence of the assembly load.

The adaptive FE simulation was used to simulate the fretting wear process for one million loading cycles. An adaptive time stepping [28] was used in the simulations with an assumption of constant wear rate during a certain number of cycles (ΔN). After several preliminary simulations, it was found that ΔN should not be assumed the same for all the periods of loading cycles. Due to the existence of a very small mismatch angle in the geometry of the interface (distal contact type with an angular mismatch of 0.01°), large contact pressures were induced over the small contacting area at the first loading cycles, which showed that care should be taken for the selection of ΔN. During the fretting wear process, the contacting area expanded gradually, which then reduced the contact pressure. Therefore, ΔN was carefully changed from 50 to 800 loading cycles over the entire fretting wear process. The size of the elements was refined several times and 0.10 mm was found as the most suitable length of the element edge in the contact area, which could provide mesh-independent results. Figure 3a shows that the first layers of the head and neck materials at the interface were meshed with very small structured quadratic elements. These elements need to be small enough to correctly model the contact pressure and relative displacement over the contact area. The sublayers were then meshed by free-quad elements, which allow increased element sizes away from the contact area. The third part of the head and neck models was again meshed by relatively large structured elements. This meshing structure considerably reduced the solution time while providing accurate results. To simulate the interaction between the head and neck, both normal and tangential contact behaviors were defined. Normal contact was simulated using a surface-to-surface contact algorithm within ABAQUS via the "hard" contact option. The tangential interaction was modelled with a classical isotropic Coulomb friction model that was implemented with a stiffness (penalty) method.

The 2D fretting wear model of the CoCr/CoCr taper junction was assembled with four different assembly forces of 2000 N, 3000 N, 4000 N, and 5000 N. A PYTHON code and a MATLAB code were developed to report the contact pressures and relative micro-motions at the contact interface, and to find the material loss in the form of worn area from the surface at various cycles (up to 1,025,000 cycles) of normal walking gait loading.

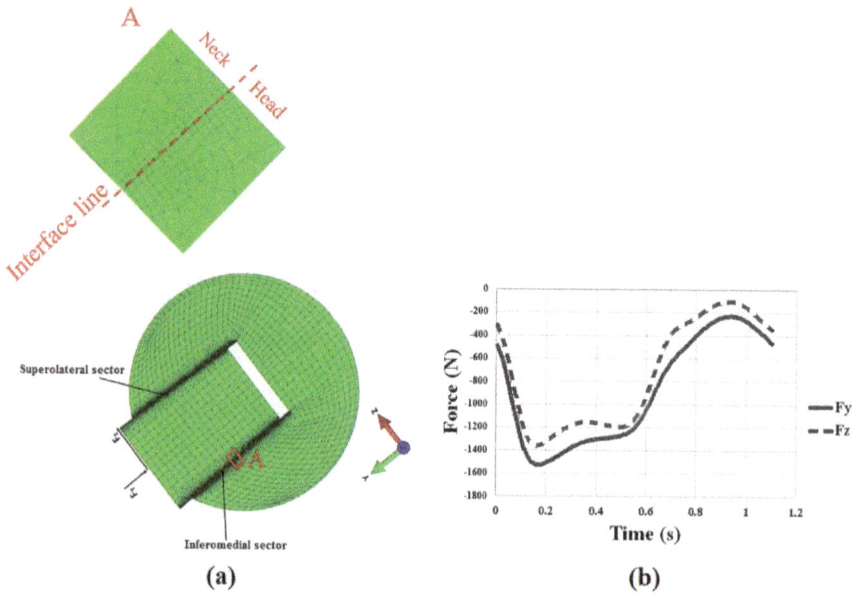

Figure 3. (**a**) Mesh structure of the 2D model of the most critical plane of the head-neck junction and (**b**) profile of corresponding forces in Y and Z axes from the normal walking gait cycle.

3. Results

3.1. Contact Pressure and Contact Length

As shown in Figure 4, when increasing the assembly force from 2000 N to 5000 N, the contact pressure increased in magnitude over the length of the neck, and the contacting region between the head and neck (contact length) also increased toward the proximal side of the neck. This confirms that a higher assembly force can further push the neck into the head, which induces greater normal contact forces. Thereby, larger contact pressures and more engagement between the head and neck surfaces (longer contact). As more loading cycles were applied (increase in the number of cycles), the peak contact pressure decreased in magnitude. The maximum magnitude of contact pressure for cases with assembly forces of 2000 N, 3000 N, 4000 N, and 5000 N decreased from 206, 257, 265, and 337 MPa at 25,000 cycles to 169, 243, 258, and 294 MPa at 1,025,000 cycles, respectively, in the super-lateral sector of the neck. These graphs can also help investigate the contact length between the head and neck. Non-zero contact stresses at any region of the surface indicate that there is contact between the head and neck in that region. After 25,000 cycles, the percentage of the neck, which is in contact with the head for cases with assembly forces of 2000 N, 3000 N, 4000 N, and 5000 N, were 48%, 64%, 75%, and 79%, respectively. These total contact lengths remained nearly constant after 1,000,000 cycles of fretting wear.

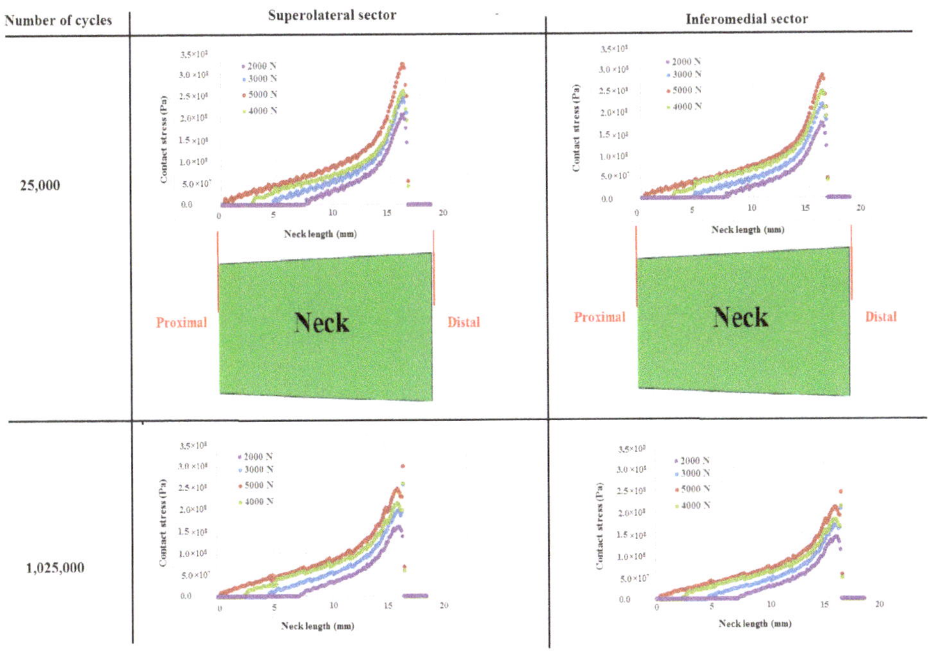

Figure 4. Variation of normal contact stress over the neck length in both super-lateral and infero-medial sectors under different assembly forces and after 25,000 and 1,025,000 loading cycles.

3.2. *Micro-Motions*

For all the assembly forces, the micro-motion at the contacting interface tends to increase from the proximal side to the distal side (Figure 5) and the magnitude of the micro-motion reduces when increasing the assembly force. The junction assembled with 2000 N had the largest micro-motions compared to the other cases with a range of 0.41 to 0.51 μm. There appears to be minimal changes in the micro-motion after 1,000,000 load cycles (Figure 5).

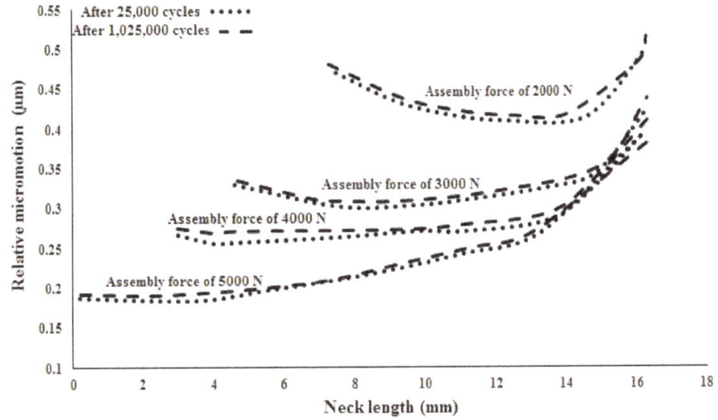

Figure 5. Relative micro-motion at the contacting interface over the neck length (super-lateral sector) for different assembly forces after 25,000 and 1,025,000 cycles.

3.3. Material Loss

Material removal over the neck length was calculated as the total area under the curve of wear depth versus the neck length in both the super-lateral and infero-medial sectors. This represents the lost area from the original edges (super-lateral and infero-medial sectors) of the 2D model. It can be seen in Figure 6a that the trend of the lost area over the number of loading cycles is linear for all of the assembly forces studied. The values of area loss for different assembly forces and at different cycles were almost equal in both the head and neck. Therefore, this figure only presents the area losses of the neck. Increasing the assembly force results in an increase in the lost area at the taper junction. For instance, when the assembly force was increased from 2000 N to 5000 N, the area loss increased from 5.28×10^{-3} mm^2 to 16.3×10^{-3} mm^2 in the neck after 1,025,000 cycles.

Figure 6b shows the effect of assembly force on the rate and location of the fretting wear damage in the form of wear depth (after 25,000 and 1,025,000 number of cycles) in the neck. It is noted that very similar depth of wear results were found in the head at the same number of cycles. These graphs can help compare the wear depth at different assembly forces, and locate the wear damage at the interface. It can be seen that the wear depths in the assembly force of 5000 N (with a maximum 0.779 μm) was significantly higher than that of the assembly force of 2000 N (with a maximum 0.413 μm).

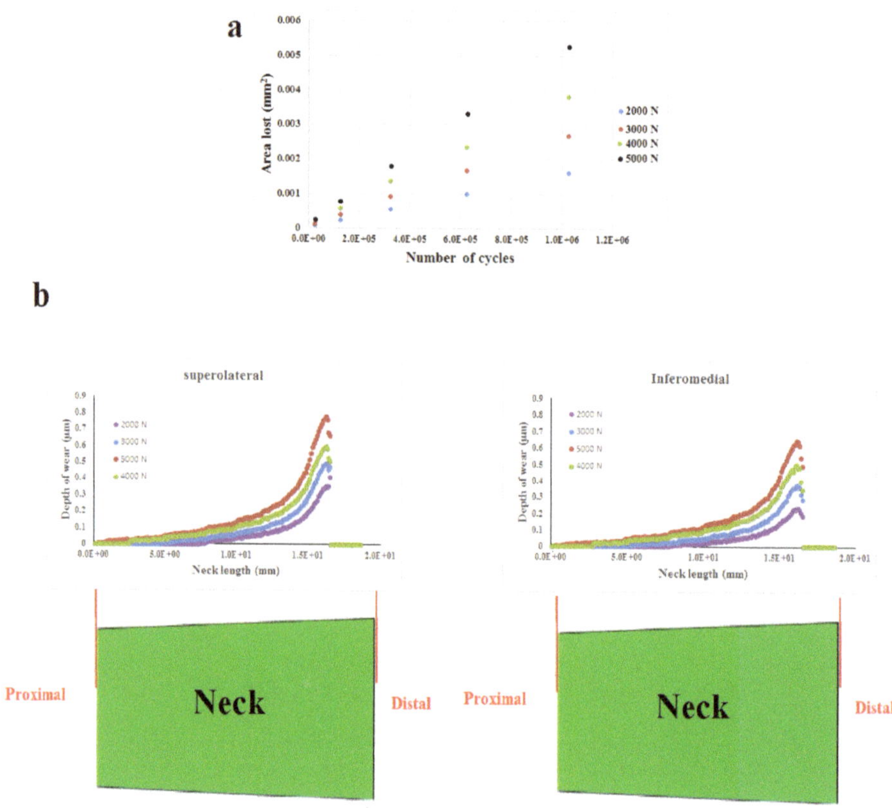

Figure 6. (**a**) Lost area versus number of cycles for different assembly forces and (**b**) depth of wear over the neck length after 1,025,000 cycles.

4. Discussion

In this work, the fretting wear mechanism and material loss were investigated in a CoCr/CoCr head-neck junction with a real angular mismatch in a PBS solution and under normal walking gait loading. The junction was assembled with various forces ranging from 2 kN to 5 kN to represent low-to-high impaction forces applied by surgeons in practice. The area loss from the edges of the most critical plane of the junction (as an indicator of material loss in the junction) showed a linearly increasing pattern over the fretting wear cycles. This could help estimate the degree of material loss after several million cycles of fretting wear.

The results of this work revealed that contact pressure, contact length, and relative micro-motion at the interface of the junction are the key parameters that can influence the material loss caused by fretting wear. Figure 5 showed that, when increasing the assembly force, relative micro-motion between the head and neck components reduces considerably, which offers more stability to the junction. According to the Archard equation, wear is proportional to both the contact pressure and relative micro-motion (amplitude of sliding). Even though the relative micro-motions decrease in the firmly assembled junctions, the significant increase in the contact pressure (induced over greater contact regions) leads to a net increase in fretting wear and, consequently, material removal. The results showed that a higher assembly force can induce a longer contact at the interface. This can extend the surface on which fretting wear is to occur and can, therefore, increase the extent of material removal. As shown in Figure 6a (for the studied taper design and material combination), increasing the assembly force results in more material loss. This is in contrast with the English's results [22] where higher assembly forces were reported to reduce fretting wear. On the other hand, Bitter's experimental results [18] showed no rational relation between the assembly force and the volumetric wear. The wear volume reduced when increasing the assembly force from 2 kN to 4 kN, and then slightly increased when increasing the assembly force from 4 kN to 15 kN. They found large standard deviations in their wear volume results (no significant difference between the three tested assembly forces). Bitter's FE simulations were too simplified and did not incorporate geometry updates to account for material loss due to the process of fretting wear. One immediate difference between the two previously mentioned studies and the present work is the material combination. In this study, the material combination is CoCr/CoCr, while they used CoCr/Ti and Ti/Ti combinations. Material properties, particularly the modulus of elasticity, can influence the behavior of the contact, especially the relative micro-motion. The angular mismatch within the junction is the major difference between the present work and their studies. The authors have previously shown that the existence of angular mismatch has a significant effect on the contact length, contact pressure, relative micro-motion, and, accordingly, the wear damage [23]. In English's model, zero mismatch was assumed between the head and neck taper angles. Therefore, the contact length would always be constant (due to having no angular mismatch).Furthermore, increasing the assembly force reduces the relative micro-motion at the head-neck interface, which, in turn, reduces the amount of material loss. However, in this work, the contact between the head and neck is not perfect. Therefore, increasing the assembly force increases the contact length in the head-neck junction, which results in increasing the material loss. Bitter's et al. [18] did not mention if there was an actual angular mismatch in their head-neck samples. Therefore, valid statements cannot be made to directly compare and discuss their results with those of this study in terms of the mismatch angle's influence.

Assembly force, as an intraoperative surgical parameter, can play an important role in the fretting wear damage to the head and neck components. This study was developed for a particular design including a CoCr/CoCr material combination with a distal angular mismatch of 0.01°. This, together with the contradicting results reported previously [18,23], may suggest that further research is required to investigate the influence of the assembly force on the fretting wear behavior, considering various angular mismatches as well as different material combinations and loading profiles of other daily activities before making a certain suggestion to clinicians in terms of a recommended force for assembling head-neck taper junctions. Moreover, fretting corrosion in the head-neck taper junction is a combination of mechanical fretting wear and electro-chemical corrosion processes. The scope of this study was to investigate the influence of assembly force (as a mechanical parameter) on the material loss at the head-neck junction of hip implants. Hence, the model was developed to simulate only the mechanical fretting wear process in the junction. However, corrosion can play an important role in the behavior of the contact and, thus, the amount of material loss. The authors of this study have developed a new adaptive finite element model to simulate fretting corrosion at metallic interfaces [30]. This model has been successfully used to simulate fretting corrosion for only a simple geometry. Further research is required to use this new and complex model to simulate fretting wear corrosion in head-neck taper junctions.

5. Conclusions

High assembly forces reduce the relative micro-motions between the head and neck at the taper junction. However, they can also increase the contact pressures and the contact region at the interface, which, in turn, may intensify the fretting wear process and, consequently, increased material removal. The results of this study showed that the effect of the last two parameters (contact pressure and contact length) was more dominant in wearing out the surface of the studied CoCr/CoCr junction with a taper angle mismatch of 0.01°. Hence, when increasing the assembly force, the degree of material loss increased for this particular design and material combination of the junction studied in this work.

Author Contributions: Conceptualization, K.F., R.H.O. and M.T.; Data curation, K.F. and H.B.; Formal analysis, K.F., R.H.O. and M.T.; Methodology, K.F., R.H.O., H.B. and M.T.; Resources, R.H.O.; Software, M.T.; Supervision, R.H.O. and M.T.; Validation, K.F. and R.H.O.; Visualization, K.F.; Writing—original draft, K.F. and R.H.O.; Writing—review & editing, R.H.O. and M.T.

Funding: This research received no external funding.

Acknowledgments: The authors acknowledge the support of an Australian Government Research Training Program Scholarship for Khosro Fallahnezhad's PhD.

Conflicts of Interest: The authors have no conflict of interest to declare.

References

1. Gilbert, J.L.; Buckley, C.A.; Jacobs, J.J. In vivo corrosion of modular hip prosthesis components in mixed and similar metal combinations. The effect of crevice, stress, motion, and alloy coupling. *J. Biomed. Mater. Res.* **1993**, *27*, 1533–1544. [CrossRef] [PubMed]
2. Oskouei, R.; Fallahnezhad, K.; Kuppusami, S. An Investigation on the Wear Resistance and Fatigue Behaviour of Ti-6Al-4V Notched Members Coated with Hydroxyapatite Coatings. *Materials* **2016**, *9*, 111. [CrossRef] [PubMed]
3. Chana, R.; Esposito, C.; Campbell, P.A.; Walter, W.K.; Walter, W.L. Mixing and matching causing taper wear: Corrosion associated with pseudotumour formation. *J. Bone Joint Surg. Br.* **2012**, *94*, 281–286. [CrossRef]
4. Fricka, K.B.; Ho, H.; Peace, W.J.; Engh, C.A., Jr. Metal-on-Metal Local Tissue Reaction Is Associated With Corrosion of the Head Taper Junction. *J. Arthroplast.* **2012**, *27*, 26–31. [CrossRef]
5. Hussenbocus, S.; Kosuge, D.; Solomon, L.B.; Howie, D.W.; Oskouei, R.H. Head-neck taper corrosion in hip arthroplasty. *BioMed Res. Int.* **2015**, *2015*, 9. [CrossRef]
6. Fallahnezhad, K.; Farhoudi, H.; Oskouei, R.H.; Taylor, M. Influence of geometry and materials on the axial and torsional strength of the head-neck taper junction in modular hip replacements: A finite element study. *J. Mech. Behav. Biomed. Mater.* **2016**, *60*, 118–126. [CrossRef] [PubMed]
7. Donaldson, F.E.; Coburn, J.C.; Siegel, K.L. Total hip arthroplasty head-neck contact mechanics: A stochastic investigation of key parameters. *J. Biomech.* **2014**, *47*, 1634–1641. [CrossRef]
8. Gührs, J.; Körner, M.; Bechstedt, M.; Krull, A.; Morlock, M.M. Stem taper mismatch has a critical effect on ceramic head fracture risk in modular hip arthroplasty. *Clin. Biomech.* **2017**, *41*, 106–110. [CrossRef] [PubMed]

9. Kocagöz, S.B.; Underwood, R.J.; Sivan, S.; Gilbert, J.L.; MacDonald, D.W.; Day, J.S.; Kurtz, S.M. Does Taper Angle Clearance Influence Fretting and Corrosion Damage at the Head-Stem Interface? A Matched Cohort Retrieval Study. *Semin. Arthroplast.* **2013**, *24*, 246–254. [CrossRef]
10. Farhoudi, H.; Fallahnezhad, K.; Oskouei, R.H.; Taylor, M. A finite element study on the mechanical response of the head-neck interface of hip implants under realistic forces and moments of daily activities: Part 1, level walking. *J. Mech. Behav. Biomed. Mater.* **2017**, *75*, 470–476. [CrossRef]
11. Fallahnezhad, K.; Farhoudi, H.; Oskouei, R.H.; Taylor, M. A finite element study on the mechanical response of the head-neck interface of hip implants under realistic forces and moments of daily activities: Part 2. *J. Mech. Behav. Biomed. Mater.* **2018**, *77*, 164–170. [CrossRef]
12. Rehmer, A.; Bishop, N.E.; Morlock, M.M. Influence of assembly procedure and material combination on the strength of the taper connection at the head-neck junction of modular hip endoprostheses. *Clin. Biomech. (Bristol, Avon)* **2012**, *27*, 77–83. [CrossRef] [PubMed]
13. Georgiou, G.; Siapkara, A.; Dimitrakopoulou, A.; Provelengios, S.; Dounis, E. Dissociation of bipolar hemiarthroplasty of the hip after dislocation. *Injury* **2006**, *37*, 162–168. [CrossRef]
14. Ramoutar, D.N.; Crosnier, E.A.; Shivji, F.; Miles, A.W.; Gill, H.S. Assessment of Head Displacement and Disassembly Force With Increasing Assembly Load at the Head/Trunnion Junction of a Total Hip Arthroplasty Prosthesis. *J. Arthroplast.* **2017**, *32*, 1675–1678. [CrossRef] [PubMed]
15. Nassutt, R.; Mollenhauer, I.; Klingbeil, K.; Henning, O.; Grundei, H. Relevance of the insertion force for the taper lock reliability of a hip stem and a ceramic femoral head. *Biomed. Tech. Biomed. Eng.* **2006**, *51*, 103–109. [CrossRef] [PubMed]
16. Pennock, A.T.; Schmidt, A.H.; Bourgeault, C.A. Morse-type tapers: Factors that may influence taper strength during total hip arthroplasty. *J. Arthroplast.* **2002**, *17*, 773–778. [CrossRef]
17. Scholl, L.; Longaray, J.; Raja, L.; Lee, R.; Faizan, A.; Herrera, L.; Thakore, M.; Nevelos, J. Friction in modern total hip arthroplasty bearings: Effect of material, design, and test methodology. *Proc. Inst. Mech. Eng. Part H J. Eng. Med.* **2015**, *230*, 50–57. [CrossRef]
18. Bitter, T.; Khan, I.; Marriott, T.; Lovelady, E.; Verdonschot, N.; Janssen, D. A combined experimental and finite element approach to analyse the fretting mechanism of the head-stem taper junction in total hip replacement. *Proc. Inst. Mech. Eng. Part H J. Eng. Med.* **2017**, *231*, 862–870. [CrossRef] [PubMed]
19. Witt, F.; Gührs, J.; Morlock, M.M.; Bishop, N.E. Quantification of the contact area at the head-stem taper interface of modular hip prostheses. *PLoS ONE* **2015**, *10*, e0135517. [CrossRef]
20. Dyrkacz, R.M.R.; Brandt, J.M.; Morrison, J.B.; O'Brien, S.T.; Ojo, O.A.; Turgeon, T.R.; Wyss, U.P. Finite element analysis of the head-neck taper interface of modular hip prostheses. *Tribol. Int.* **2015**, *91*, 206–213. [CrossRef]
21. English, R.; Ashkanfar, A.; Rothwell, G. A computational approach to fretting wear prediction at the head–stem taper junction of total hip replacements. *Wear* **2015**, *338–339*, 210–220. [CrossRef]
22. English, R.; Ashkanfar, A.; Rothwell, G. The effect of different assembly loads on taper junction fretting wear in total hip replacements. *Tribol. Int.* **2016**, *95*, 199–210. [CrossRef]
23. Fallahnezhad, K.; Oskouei, R.H.; Badnava, H.; Taylor, M. An adaptive finite element simulation of fretting wear damage at the head-neck taper junction of total hip replacement: The role of taper angle mismatch. *J. Mech. Behav. Biomed. Mater.* **2017**, *75*, 58–67. [CrossRef] [PubMed]
24. Farhoudi, H.; Oskouei, R.H.; Jones, C.F.; Taylor, M. A novel analytical approach for determining the frictional moments and torques acting on modular femoral components in total hip replacements. *J. Biomech.* **2015**, *48*, 976–983. [CrossRef] [PubMed]
25. Farhoudi, H.; Oskouei, R.; Pasha Zanoosi, A.; Jones, C.; Taylor, M. An Analytical Calculation of Frictional and Bending Moments at the Head-Neck Interface of Hip Joint Implants during Different Physiological Activities. *Materials* **2016**, *9*, 982. [CrossRef]
26. Archard, J. Contact and rubbing of flat surfaces. *J. Appl. Phys.* **1953**, *24*, 981–988. [CrossRef]
27. Johansson, L. Numerical Simulation of Contact Pressure Evolution in Fretting. *J. Tribol.* **1994**, *116*, 247–254. [CrossRef]
28. Ding, J.; Leen, S.B.; McColl, I.R. The effect of slip regime on fretting wear-induced stress evolution. *Int. J. Fatigue* **2004**, *26*, 521–531. [CrossRef]
29. McColl, I.R.; Ding, J.; Leen, S.B. Finite element simulation and experimental validation of fretting wear. *Wear* **2004**, *256*, 1114–1127. [CrossRef]

30. Fallahnezhad, K.; Oskouei, R.H.; Taylor, M. Development of a fretting corrosion model for metallic interfaces using adaptive finite element analysis. *Finite Elem. Anal. Des.* **2018**, *148*, 38–47. [CrossRef]
31. Maruyama, N.; Kawasaki, H.; Yamamoto, A.; Hiromoto, S.; Imai, H.; Hanawa, T. Friction-wear properties of nickel-free co-cr-mo alloy in a simulated body fluid. *Mater. Trans.* **2005**, *46*, 1588–1592. [CrossRef]

metals

MDPI

Review

3D Printed Acetabular Cups for Total Hip Arthroplasty: A Review Article

Lorenzo Dall'Ava *, Harry Hothi, Anna Di Laura, Johann Henckel and Alister Hart

Institute of Orthopaedics and Musculoskeletal Science, University College London and the Royal National Orthopaedic Hospital, Stanmore HA7 4LP, UK
* Correspondence: lorenzo.dallava.17@ucl.ac.uk

Received: 31 May 2019; Accepted: 26 June 2019; Published: 28 June 2019

Abstract: Three-dimensional (3D) printed titanium orthopaedic implants have recently revolutionized the treatment of massive bone defects in the pelvis, and we are on the verge of a change from conventional to 3D printed manufacture for the mass production of millions of off-the-shelf (non-personalized) implants. The process of 3D printing has many adjustable variables, which taken together with the possible variation in designs that can be printed, has created even more possible variables in the final product that must be understood if we are to predict the performance and safety of 3D printed implants. We critically reviewed the clinical use of 3D printing in orthopaedics, focusing on cementless acetabular components used in total hip arthroplasty. We defined the clinical and engineering rationale of 3D printed acetabular cups, summarized the key variables involved in the manufacturing process that influence the properties of the final parts, together with the main limitations of this technology, and created a classification according to end-use application to help explain the controversial and topical issues. Whilst early clinical outcomes related to 3D printed cups have been promising, in-depth robust investigations are needed, partly because regulatory approval systems have not fully adapted to the change in technology. Analysis of both pristine and retrieved cups, together with long-term clinical outcomes, will help the transition to 3D printing to be managed safely.

Keywords: 3D printing; additive manufacturing; orthopaedic implants; total hip arthroplasty; acetabular cups

---

1. Introduction

Three-dimensional (3D) printing, also known as 'Additive Manufacturing' (AM), of titanium orthopaedic implants has revolutionized the treatment of massive bone defects in the pelvis due to their ability to be customized with complex shapes, size and surface geometries; this is more complicated to achieve with conventional manufacturing (i.e., non-3D printing) methods, such as drop forging and machining, which are commonly used to produce orthopaedic implants. The greatest impact of 3D printing of orthopaedic implants is, however, still to come: the mass production of millions of off-the-shelf (non-personalized) implants.

Every year, over 600,000 total hip arthroplasty (THA) procedures are performed in Europe and 1.4 million worldwide; these numbers are expected to grow significantly by 2030 [1–3]. The main clinical rationale for the use of 3D printed off-the-shelf implants is achieving a successful long-term fixation with bone to restore biomechanical function of the joint.

The process of 3D printing has many adjustable variables which, taken together with the possible variation in designs that can be printed, has created even more variables in the final product that must be understood if we are to predict the safety and performance of 3D printed implants [4–7]. The regulatory approval systems have not yet completely caught up with the change in technology [8];

surgeons prefer to use implants that have been followed up for several years and have been highly rated by systems such as the Orthopaedic Data Evaluation Panel (ODEP, UK) [9]. Orthopaedics has already shown cases of unpredictable outcomes with design solutions that were thought to be revolutionary, such as metal-on-metal hip replacements [10,11].

This review aims to describe the role of 3D printing of orthopaedic implants, focusing on acetabular components used in THA. To achieve this, we (1) explain the rationale for 3D printed acetabular cups, (2) describe the variables and the limitations involved in the 3D printing manufacture, and (3) suggest a classification for these cups, presenting also the investigation methods and the clinical outcomes of 3D printed cups.

2. Rationale for 3D Printing in Orthopaedics

Although the majority of THA procedures are still performed using conventionally manufactured cups, acetabular components produced using 3D printing technologies are being increasingly used for primary and revision hip surgeries. There are advantages and disadvantages of each production techniques (Table 1). In terms of manufacturing, 3D printing enables complex porous structures with specifically designed pores shapes to be produced, unlike conventional technologies, where the control over the final architecture of the porous backside coating is limited. Furthermore, customization of implants can be achieved more easily using 3D printing. In terms of clinical outcomes and investigation of the implants, the conventional cups have a long-track record, with long-term clinical results, unlike 3D printed; however, aseptic loosening (i.e., loss of fixation without infection) is still one of the most common reasons for revision [12]. Independent investigations of full-post production 3D printed acetabular components are currently missing.

Table 1. Summary of advantages and disadvantages of 3D printing and conventional manufacturing for acetabular components in hip arthroplasty [6,7,12,13].

Factor	3D Printing	Conventional Manufacturing
Advantages	• Complex and easily adjusted porous structure for enhanced fixation • Cup size optimization • Easily customized/personalized	• Widespread clinical use • Long-term clinical outcomes
Disadvantages	• Potential risks and clinical impact poorly understood • Absence of implants investigations • Few reported clinical outcomes	• Poor fixation still an issue in THA • Limited design freedom • Customization limited

2.1. Clinical Rationale for 3D Printed Cups

Customized implants

The clinical unmet need addressed by customized 3D printed titanium acetabular implants was the poor outcome resulting from the use of conventionally manufactured triflange implants, jumbo cup, cages and augments to reconstruct massive acetabular defects [14–18]. These defects most commonly occur following previous failed implants. Conventionally manufactured implants failed due to poor fixation to the bone as a result of the complex shape of the defect and low surface area of host bone with good blood supply. 3D printed implants have overcome both of these problems (Figure 1a).

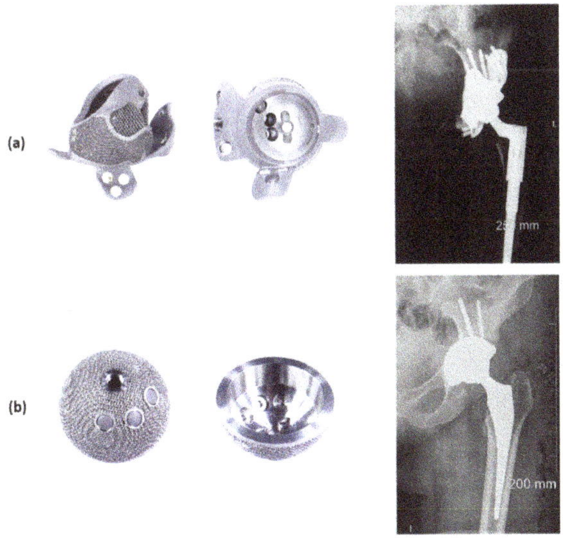

Figure 1. Images showing backside and internal surface of (**a**) custom and (**b**) off-the-shelf 3D printed implants. Post-operative X-ray images are also shown.

Off-the-shelf

One of the commonest reasons for the failure of orthopaedic implants is loosening from the bone [19]. Several factors are responsible for this including poor bone quality, the presence of metabolic bone diseases, pathological bone anatomy or conditions due to previous surgeries [20,21]. 3D printing creates implants with highly porous structures for enhanced fixation, matching bone characteristics such as pore size, coefficient of friction and modulus of elasticity to avoid stress shielding. Porosity and properties similar to cancellous bone may enable bone-implant interactions that lead to primary stability, bone ingrowth and osseointegration [22].

3D printed, off-the-shelf implants can also be designed to optimize hip joint biomechanics and cup size. Surgeons try to use the 36 mm heads for greater stability [23,24], but this is difficult to achieve with conventional implants in a size suitable for most women unless they can be made with thinner walls. 3D printing can achieve this, allowing surgeons to use large femoral head size options (e.g., 36 mm diameter) with smaller acetabular cups (e.g., 48 mm diameter). It is not clear, however, what impact this design modification may have on the mechanical properties of the cup (Figure 1b).

2.2. Engineering Rationale for 3D Printed Cups

The ISO/ASTM 52900 standard defines 3D printing as 'the process of joining materials to make parts from 3D model data, usually layer upon layer, as opposed to subtractive manufacturing and formative manufacturing methodologies' [25]. From an engineering perspective, the greatest advantage of 3D printing, when compared to conventional techniques, is the design freedom enabled during the Computer Aided Design (CAD) process. Unlike conventional methods, tools such as molds are not needed, reducing the cost of the final part; therefore, increased complexity in the structure of the 3D printed object does not involve higher costs.

To date, 3D printing involves a lower cost-per-part than conventional manufacturing when the number of parts is below a certain threshold, but with the increased adoption of this manufacturing technique, this threshold will grow [6]. It has also been estimated that the consumption of raw materials may be reduced up to 75% [26]. Acetabular cups with complex architecture and controlled mesostructure (structure associated with pores and porosity) can be designed and made, integrating porous and dense regions.

3. 3D Printing Manufacturing Process, Limitations and Risks

Both conventional and 3D printing technology use Titanium-6-Aluminum-4-Vanadium (Ti6Al4V) alloy, due to its biocompatibility, mechanical strength and corrosion resistance [27,28]. Guidelines regarding the properties of Ti6Al4V as a material for surgical implants are defined by both the International Organization for Standardization (ISO) and the American Society for Testing and Material (ASTM) standards [29–31]. However, we must be aware of the limitations and risks of defects that are unique to the 3D printing process, together with the potential impacts of these when present in orthopaedic implants.

3.1. Manufacturing Process

Conventional manufacturing technologies start production of the implant from a dense block that is machined into the shape of an acetabular component, using techniques such as computer numerical controlled (CNC) machining of wrought or cast bars, powder metallurgy, drop forging and casting [27,32]. The component is then finished using wet or coarse blasting on the external surface to obtain roughened areas, and post-processing the internal surface to obtain the required dimensional tolerances and minimal friction for an optimal seating of the liner [27]. The backside coating is applied at a later step using different methods such as solid state processing (powder metallurgy, sintering of powders and fibres), vapour deposition or plasma spray [22,33–35].

The 3D printing, or additive manufacturing, methods used to manufacture acetabular cups are classified as Powder Bed Fusion (PBF) technologies, where an energy source (laser or electron beam) selectively melts specific regions of a powder bed. Two different processes fall under the PBF category: selective laser melting (SLM) [36,37] and electron beam melting (EBM) [7,38–41]. In the pastdecade 3D printing has evolved, leading in 2007 to the first acetabular component produced using EBM to obtain the CE-certification (Figure 2) [42]. To date, several companies have commercialized machines with laser powder-bed hardware (EOS GmbH, Kraillin, Germany; SLM Solutions, Lübeck, Germany; Concept Laser, GE Additive, Lichtenfels, Germany; Renishaw, Wotton-under-Edge, UK; 3D Systems, Rock Hill, SC, USA) and one company (Arcam, GE Additive, Mölndal, Sweden) with electron beam as the energy source [6,43,44].

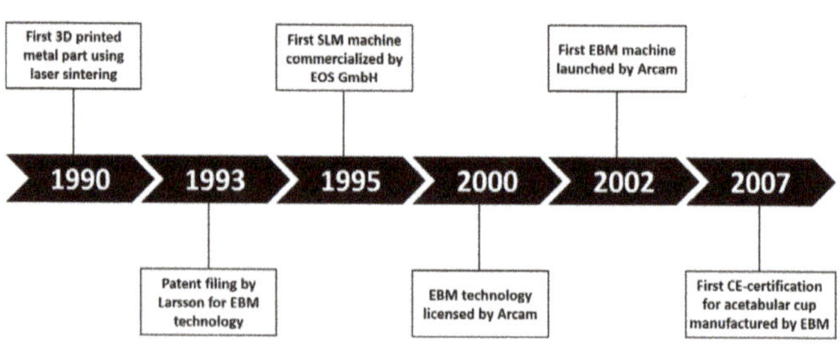

Figure 2. Timeline chart of the evolution of 3D printing in orthopaedics (EBM, electron beam melting; SLM, selective laser melting).

These 3D printing processes are guided by CAD files containing the model of the part to manufacture. With the layer-over-layer process, once a layer of powder has been selectively melt, the build platform is lowered, new powder is deposited and raked, and the process is repeated until the object is built [45]. Adherence of the current layer to the rest of the part is achieved by re-melting of previous layers. The whole implant (dense and porous parts) is produces in a single step, although post-processing such as powder removal, heat treatment or post-machining are required [4,6,37,43,44].

It has been estimated that more than 130 variables are involved in PBF [46,47]; a schematic summary of some of these variables is presented in Figure 3.

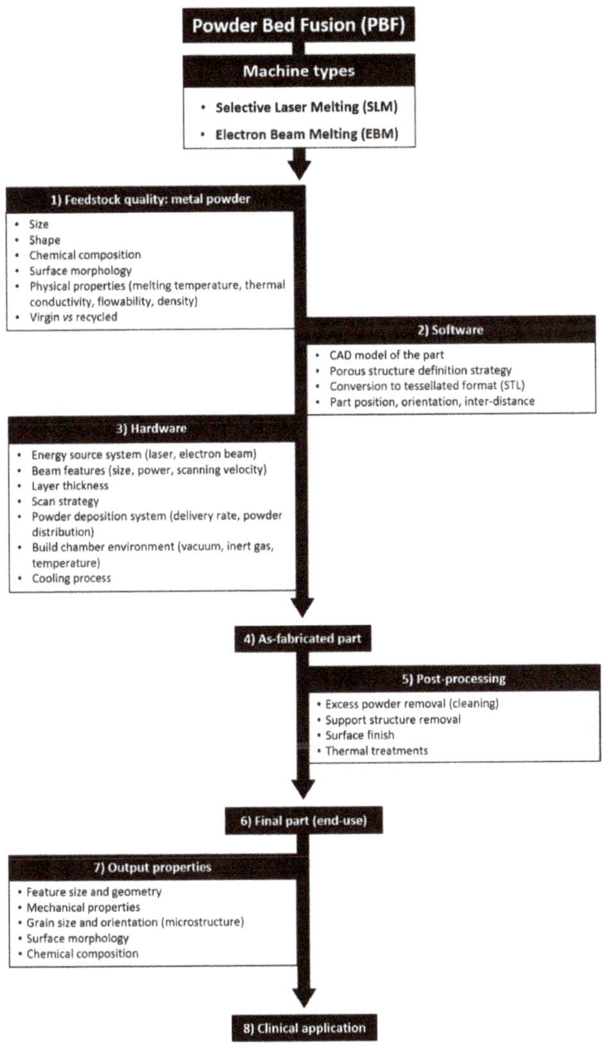

Figure 3. Flowchart of the powder bed fusion process showing the main variables involved. The output properties of the final part are determined by feedstock quality (metal powder), software and hardware specifics, and post-processing.

The feedstock quality is the first 'macro-variable' to determine the properties of the final implant. Powder characteristics, such as size, shape, chemical composition and surface morphology depend on the powder production technique (gas atomization, rotary atomization, plasma rotating electrode process, plasma atomization); these influence properties like flowability (how well a powder flows), apparent density (how well a powder packs) and thermal conductivity of the whole powder bed [43,44]. Smaller powder particles are used in SLM, compared to EBM, leading to a smoother surface finish due to decrease of the size of satellites formed during melting and reduction of layer thickness; this has an effect on the overall roughness, which is higher in EBM-manufactured components [44]. Moreover, powder chemical composition must be within the alloy specifications, especially when the material is recycled and the possibility of contamination with oxygen or other gases is present. It has been demonstrated that oxygen content can increase up to 0.19% in weight over 21 reuse cycles [48], implying that powder can be reused, but titanium and its alloys are known to oxidize and suffer embrittlement when the oxygen content increases [49].

The second 'macro-variable' is defined by the software system; the CAD model represents the ideal final shape of the part to build, but it has to be converted into a surface tessellation format (STL) and sliced into layers in order to generate the manufacturing information for each cross-section. The conversion CAD-to-STL may represent a source of error with loss of resolution because the STL format simplifies the component geometry into a set of triangular facets connected at the vertices [4,47,50]. Moreover, the spatial position and the orientation of the components in the build chamber may influence the geometrical accuracy of the part, because different temperature gradients can result in the powder bed [51].

The third 'macro-variable' is represented by the hardware of the machine. SLM works using a laser source, a system of lenses and a galvanometer (scanning mirror) to position the beam; the laser heats and melts the powder when photons are absorbed by the powder particles (Figure 4a). EBM uses a filament (usually made of tungsten) as source of electrons, a magnetic coil to collimate and deflect the beam spatially and a column for the electron beam, resembling a high-power scanning electron microscopy (SEM) (Figure 4b). In this case, the powder is heated by the transfer of kinetic energy from incoming electrons, generating also increasingly negative charge on the powder bed. In order to reduce this, the EBM energy is more diffuse (i.e., larger heat-affected powder zone) and helium gas is injected during melting to dissipate the charge; this leads to a larger minimum feature size that can be produced and to the use of larger powder particle size [34,43–45]. In general, the small feature size that both SLM and EBM can manufacture allows to print the meshes or foam structures that are present in orthopaedic implants such as acetabular cups [52].

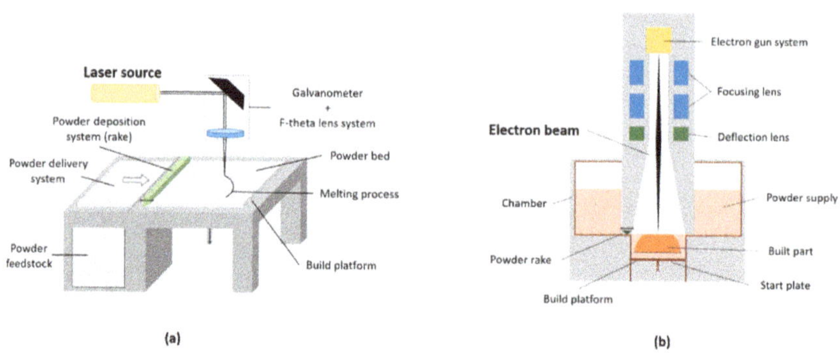

Figure 4. Schematic representations of the two-powder bed fusion machines used to manufacture orthopaedic implants: (**a**) selective laser melting (SLM) and (**b**) electron beam melting (EBM). The two technologies use the same powder-bed principle for layer-by-layer selective melting, but hardware and process differences are present.

Other variables to be considered in the 3D printing processes can be summarized as follows: scan speed, which is higher in EBM because magnetically driven, while in SLM depends on the galvanometer inertia; layer thickness, which determines how much powder is distributed to the melt surface and is greater in EBM because of larger powder particles size and beam focusing; powder deposition, which is delivered by a powder hopper and a metal rake in EBM and by variable feeding systems (hopper or dispersing piston) and re-coating systems (soft blades or roller) in SLM; build chamber atmosphere, which is filled with inert gas (argon or nitrogen) to avoid oxidation in SLM and is under vacuum ($<5 \times 10^{-4}$ Pa) to generate the electrons and prevent any contamination in EBM [43,44].

A fundamental step of the building process is the cooling phase, which determines both microstructural and mechanical properties of the 3D printed part. This process is influenced by parameters such as scan strategy (path followed by the heating source) and temperature of the powder bed. The higher energy input of the electron beam heats the surrounding loose powder, generating a higher temperature in the chamber (400–1000 °C), compared to SLM (100–200 °C), where heaters are used to avoid temperature drops on the build platform. Therefore, the thermal cycling experienced by the metal (simultaneous melting of the top powder, re-melting of underlying layers and cooling) is different between the two PBF techniques, resulting in different microstructure (grain size and orientation, phase distribution). Laser-built components experience the presence of residual stresses which must be relieved [4,43,44]. A summary of the differences between SLM and EBM is presented in Table 2.

At the end of the building process, the part is considered "as-fabricated" and requires post-processing steps. Excess powder and support structures must be removed; thermal treatment can be applied to reduce residual stresses in the structure and enhance the overall mechanical properties. Machining can also be used to modify the surface finish, influencing the actual part tolerance, minimum feature size and surface roughness of the end-use component. As example, surface roughness is influenced by the layering effect ('stair-step effect'), by the actual roughness of the powder particles (finer powder means smaller satellites formation) and by post-processing steps. Similarly, the geometrical accuracy depends on the setup of the machine and post-processing treatments [43,44]; it has been suggested that EBM tolerance capability is about ±250/300 μm [53,54].

The mechanical properties of the final part depend on the microstructure that is formed during the building process, which in turn depends on the parameters of the technology [45].

At room temperature, the Ti6Al4V alloy is made of α- and β-phases; after being processed using EBM or SLM, different microstructures are generated [43]. The low cooling rate experienced during EBM enables the β-phase formed during the process to transform back to α-phase, whilst the high cooling rate of SLM generates martensitic α′-phase. This is usually removed using post-processing heat treatments. Overall, the thermal history (melting and cooling), the quality of the feedstock, the potential presence of structural defects, the component size, the energy source density and the scanning strategy are the main factors affecting the final mechanical properties [55–57]. The layer-over-layer building process also creates anisotropy in the build direction (i.e., Z-direction).

Table 2. Main parameters of the main parameters of selective laser melting (SLM) and electron beam melting (EBM) [37,42–44,58–60].

Features	SLM	EBM
Heat source	Laser beam (up to 1 kW)	Electron beam (60 kW)
Scan speed	Limited by galvanometer inertia	Fast, magnetically driven
Powder size	10–45 μm	45–106 μm
Minimum beam size	50 mm	140 mm
Beam/melt pool dimension	0.5–1.5 μm	2–3 μm
Layer thickness	20–100 μm	50–200 μm
Chamber atmosphere	Argon or nitrogen	Vacuum (+helium)
Environment temperature	Build platform at 100–200 °C	Chamber at 400–1000 °C
Powder pre-heating	Using infrared or resistive heaters	Using electron beam
Surface finish	Excellent to moderate (~20 μm)	Moderate to poor (~35 μm)
Residual stresses	Yes	No

If the 3D printed part also includes porous structures, as with acetabular cups, the mechanical properties of the whole component will differ and depend on the behaviour of this region, as demonstrated by Murr et al. [13]

3.2. *Limitations and Potential Risks of 3D Printing*

Dimensional accuracy, optimal surface roughness and minimization of residual stresses in the material structure are key aspects of 3D printed parts for critical applications like medical implants. These can be achieved choosing the most adequate powder feedstock, correctly designing the final object and selecting the best machine parameters. However, 3D printing shows limitations that leads to the presence of defects in the built part.

One of the most common limitation is represented by the presence of voids or pores in the structure of the final part, which can affect its mechanical properties [45]. These defects can be due to the properties of the powder feedstock or to suboptimal building conditions. Spherical gas pores entrapped in the powder particles can be transferred to the final object if the powder is not completely melted; similarly, gasses of the build chamber can be entrapped in the melt pool during the process, resulting in spherical voids [61]. Another source of defects is the so-called 'keyhole melting' mode, where the depth of the melt pool is controlled by the evaporation of the metal due to the high-power energy source. If the vapor cavity collapses, then almost spherical voids are left behind [62]. Voids with elongated shapes can also be generated if a lack of fusion between subsequent layers occurs (lack of fusion defects) [63].

Another common limitation is given by the presence of solid powder particles partially attached to the surface of the built part (Figure 5); if the energy source is too low, then the powder is not completely molten, whilst if it is too high, the molten pool can form small 'islands' that generate particles ('balling phenomenon') [45].

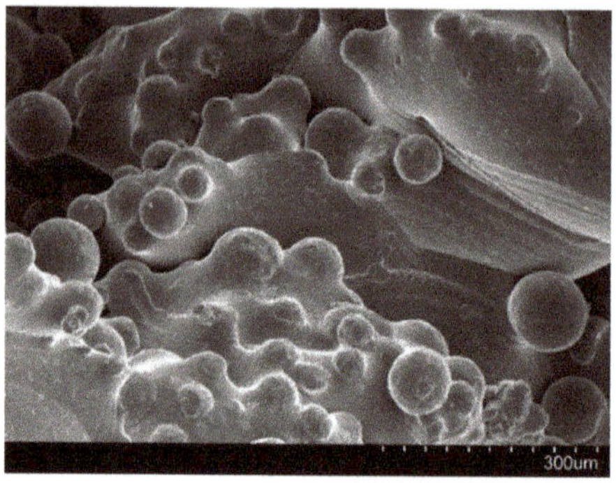

Figure 5. Scanning electron image (×150) taken at our laboratory showing a 3D printed part with presence of partially molten powder particles attached to the surface of the object. The part was manufactured using electron beam melting.

Cracking and delamination are other possible defects exhibited by 3D printed parts; these can occur due to solidification shrinkage (thermal contraction) of the structure, which subsequently generates tensile stresses that form cracks at the grain boundaries if the strength of the material is exceeded. When the residual stresses overcome the strength of the metal at the interface of subsequent layers, then a delamination phenomenon of these consecutive layers can occur [45]. When the temperature of the melting material is too elevated, a phenomenon of loss of alloying elements can also occur. This depends on the metal alloy and can cause modification in the alloy composition [64].

At the end of the manufacturing process, the as-fabricated part needs to be cleaned from the residual unmolten powder [43]. The complex porous structures that can be produced with 3D printing have made this cleaning process challenging [35,65,66].

Although all these limitations and the presence of defects have been widely reported in 3D printed parts, few studies have addressed this for medical implants. In a comprehensive review on additive manufacturing of medical instruments, Culmone et al. [66] highlighted how the presence of debris in the final component is still a technical issue that needs to be addressed. Another study by du Plessis et al. [67] identified pores in the material structure of a nasal cavity implant made of Ti6Al4V produced using SLM; these pores were in the range 10–60 μm, according to the region analyzed. In general, an issue found with 3D printed device related to surgery, such as anatomic models and surgical guides, is the non-satisfactory accuracy of the final components compared to the initial model [68].

It cannot be excluded that some of these defects are present in 3D printed acetabular cups, although no studies have reported this. It can be speculated that if some defects were to be found, then, there might be implications for the implant's properties and performance. As an example, the presence of unmolten particles on the surface of the cups might lead to the release of titanium.

To date, 3D printed implants have received both the Food & Drug Administration (FDA) and CE marking approval. In the United Kingdom, three 3D printed acetabular cup designs are currently going through the Beyond Compliance initiative, which monitors new or modified implants placed on the market. The substantial equivalence to other legally marketed devices, regarding safety and effectiveness, is still applied as a principle to clear implants, including 3D printed cups. These components are treated as conventionally manufactured implants, since the intended use (i.e., acetabular component for THA) is the same [69].

4. Classification and Analysis of 3D Printed Acetabular Cups

Over fifteen 3D printed acetabular cup designs are currently commercially available (Table 3); this number has grown constantly since 2007, when the first cups were released (Figure 6). It has been estimated that more than 60,000 acetabular cups produced using EBM have been used worldwide by 2017 and this number is expected to grow [39].

Table 3. 3D printed acetabular cup designs currently on the market (custom designs are specified) [65,70–80].

Company	Cup Brand	Porous Structure	Properties: Porosity; Pore Size
Adler Ortho (Milan, IT)	Agilis Ti-Por® *Custom* Fixa Ti-Por® Omnia Ti-Por® PolyMax Ti-Por®	Tri-Por™Cup	65%; 700 μm
Corin (Cirencester, UK)	Trinity™ Plus	Porous Layer Unique Structure (PLUS)	50–90%; 300–900 μm
Implantcast (Buxtehude, DE)	C-Fit 3D® (custom) EcoFit®	EPORE®	60%; 100–500 μm
Lima Corporate (Udine, IT)	Delta TT Delta ONE TT Delta Revision TT Promade (custom)	Trabecular Titanium (TT)	65%; 640 μm
Materialise (Leuven, BE)	aMace® (custom)	aMace®	70%; 720 μm
Medacta (Castel San Pietro, CH)	Mpact®	3D Metal™	~75%; 600–800 μm
Smith&Nephew (Memphis, USA)	Redapt	Conceloc Advanced Porous Titanium	~67%; 202–934μm
Stryker (Mahwah, USA)	Trident® II	Tritanium® AMagine™	55–65%; 100–700 μm
Zimmer Biomet (Warsaw, USA)	G7	OsseoTi™	~70%; 475 μm

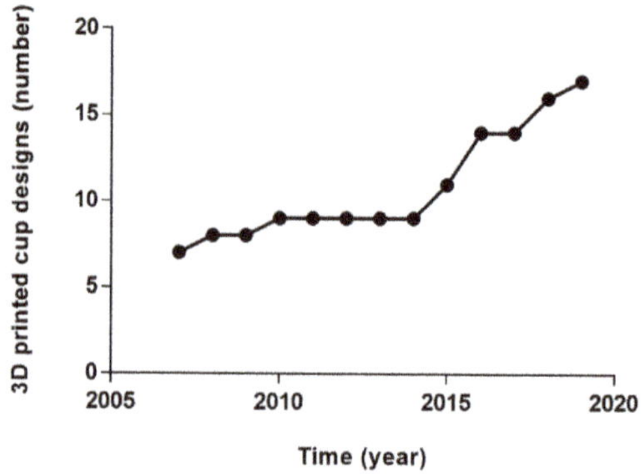

Figure 6. Graph showing the number of acetabular cup designs produced using 3D printing placed on the market in the past decade. In 2007 there were six available designs, in 2019 the number has reached seventeen, data from [70–81].

According to their end-use application, the cups can be classified as 'custom' or 'off-the-shelf'. Considering the design of the porous structure present on the cups, another classification for 3D printed acetabular components can be suggested.

4.1. Acetabular Cup Classification: Porous Structure Design

Regardless of the end-use application, another classification may take into account the different porous architectures that each company have developed. The overall aim is to obtain an open-pore (cellular or lattice) structure that can promote bone ingrowth and prevent a later-stage loosening of the implant, avoiding micromotion over 150 μm [82]. Moreover, the porous structure can reduce the risk of stress shielding due to stiffness mismatch between the metallic implant and bone, providing a lower modulus of elasticity [13,83,84]. Similarly to implants with conventionally manufactured porous coatings, the fixation efficacy of 3D printed components strongly depends on the geometric characteristic of the porous structure [33]. A compromise in maintaining the mechanical strength while providing adequate pore size leads to different porous architectures. The majority of the specifications are proprietary but the possible methods to define the 3D porous structures are well established [34,47,85]. Regular (repeated unit cells) or irregular (stochastic, random) structures can be designed and manufactured, theoretically producing any kind of porous shape; typical design strategy for periodic regular networks include the use of CAD tools (libraries of unit cells), image-based structure taken from computed tomography (CT) or magnetic resonance imaging (MRI) data or implicit surfaces solutions, where a single mathematical equation defines the pore shape [34,85]. These solutions are usually based on a trial-and-error approach, which is why the so-called 'topology optimization' have been applied in order to create structures with desired properties satisfying prescribed constraints. These techniques have been thoroughly described elsewhere [34]. In general, the quality and the real dimensions of the mesh may vary significantly depending on design and fabrication process.

An example of a unit cell-based highly porous trabecular structure is represented by Trabecular Titanium (TT; Lima Corporate, Udine, Italy): a hexagon-shaped structure developed on three planes which mimics a diamond crystal (Figure 7a) [53,54,86,87]. Differently, a stochastic architecture is represented by Tritanium (Stryker, Mahwah, NJ, USA), which is a random interconnected structure with irregular pore size and shape (Figure 7b). An example of porous structure derived from human CT data is represented by OsseoTi (Zimmer Biomet, Warsaw, IN, USA).

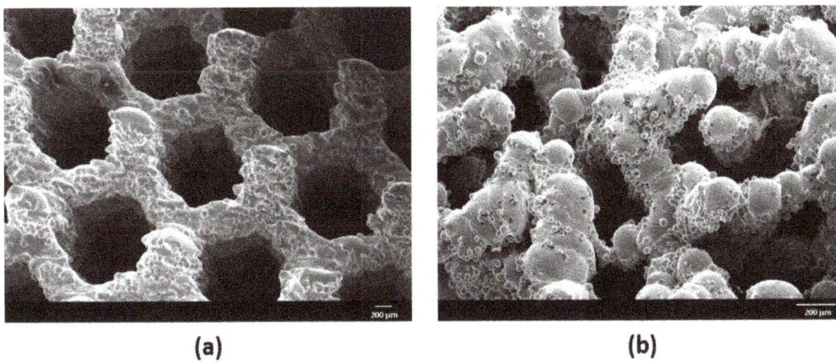

Figure 7. Scanning electron microscopy images taken at our laboratory showing two examples of 3D printed porous structures: (**a**) regular shaped architecture (Trabecular Titanium, TT; Lima Corporate, Udine, Italy) and (**b**) irregular shaped architecture (Tritanium; Stryker, Mahwah, NJ, USA). The first one is produced using EBM, the second one SLM.

4.2. Acetabular Cup Analysis

The number of investigations of 3D printed acetabular designs is limited. A summary of the analysis methods used for these cups is reported in Table 4.

Table 4. Summary of the investigation performed on 3D printed acetabular cups.

Authors	Acetabular Cup	Investigation	Findings
Davignon et al. [88]	Trident II Tritanium	Comparative analysis with conventional cup of seating and initial stability in bench test with foam blocks	3D printed cup showed significantly higher yield moment (i.e., failure of fixation) and lower seating
Kourra et al. [89]	Prototype cup produced using EBM Arcam machine	Porous structure analysis using micro-computed tomography (micro-CT)	Micro-CT can accurately provide qualitative and quantitative data on porous structure of 3D printed cups
Jahnke et al. [90]	Ecofit EPORE	Comparative analysis with two conventional cups of primary stability and relative micromotion in simulator test with foam blocks	3D printed cup showed similar micromotion to conventional cups, within an acceptable clinical range
Le Cann et al. [91]	Unspecified cup by Alder Ortho	Surface roughness influence on primary stability in bench test with foam block and animal bone	Macro-roughness is good for osseointegration but has negative effects on primary stability if present in the equatorial region of the cup

Other investigations tools have been suggested to assess the properties of 3D printed samples, such as electron microscopy, mechanical testing and histology. However, these methods have been applied to cylindrical or cubic specimens to test the properties of the structure, evaluate bone ingrowth and the interaction with bone cells [13,32,40,52,53,92–94].

Non-clinical tests, defined as 'performance tests' in the FDA 510(k) reports, have been performed on 3D printed cups in order to determine the substantial equivalence to conventionally manufactured implants and obtain the certification. Shear testing, both static and fatigue, as well as tensile testing, deformation testing, locking mechanism strength, range of motion testing, dynamic compression, fatigue testing, push-out, lever-out and torque-out testing have been carried out based on standards [95–97]. However, the outcomes of such tests are not publicly available.

To date, a small number of clinical studies about 3D printed cups have been published for both off-the-shelf [73,98–102] and custom implants [21,78,103–108].

A registry study by Castagnini et al. [73] compared the survival rate and reason for revision of a specific 3D printed off-the-shelf design (Fixa Ti-Por; Adler Ortho, Milan, Italy) with all the other cementless cups used from July 2007 and December 2015 reported in the Registry of Prosthetics Orthopaedic Implants (RIPO). The AM socket was implanted in 9,864 cases out of 36,787, with similar patients mean age at the time of surgery (68 vs. 69 years). Significantly higher survival rate (98.7% vs. 97.9%) and lower incidence of aseptic loosening were reported for the 3D printed socket. Steno et al. [98] reported on the early outcomes (mean follow-up period of 38.14 months) of three AM designs (Delta TT, Delta ONE TT, Delta Revision TT; Lima Corporate, Udine, Italy) used in 81 revision cases (mean age 68 years) between March 2009 and May 2012. Only one re-revision was performed due to instability of the acetabular component; in general, encouraging results were showed. Perticarini et al. [99] reported good clinical and radiographic mid-term results (mean follow-up period of 72.7 months) of the Delta TT cup (Lima Corporate, Udine, Italy) used in 134 THA and 8 revisions carried out between September 2007 and November 2009 (mean age 57.5 years). At the last follow-up, both improvement of the Harris Hip Score (HHS) [109] and radiographically stable acetabular components (99.3%) were recorded; hip dislocation occurred in two cases, aseptic loosening in one patient. Similar clinical and radiographic outcomes were reported by Gallart et al. [101] regarding the use of Delta TT and Delta Revision TT (Lima Corporate, Udine, Italy) in 67 revision and 5 primary cases performed between January 2009 and July 2014 (mean follow-up 30.5 months). Aseptic loosening (2 cases), infection (3 cases) and dislocation (3 cases) were recorder as reason for re-revision. Massari et al. [100] evaluated the osseointegration of Delta TT cups (Lima Corporate, Udine, Italy) implanted in 91 primary THA between February 2009 and November 2010 in four centres. Periacetabular changes in bone mineral density (BMD) were evaluated using dual-emission X-ray absorptiometry (DEXA) at fixed time points after surgery. An initial decrease in BMD was noted, followed by a progressive increase and stabilization at the last follow-up (24 months), in accordance with previous studies related to cementless acetabular components. Significant improvements in HHS and good radiographic outcomes were also reported, with no revision needed. Short and mid-term (mean follow-up of 58.6 months) clinical and radiographic results of the AM cups Delta One TT and Delta Revision TT (Lima Corporate, Udine, Italy) were also described by De Meo et al. [102]. Among the 154 revision procedures (mean age 69.6 years) performed between December 2008 and August 2015, re-revision was required in 3 cases due to dislocation. In general, more consistent conclusion about the performance of 3D printed off-the-shelf implants may be reached with long-term clinical follow-up, which are still unpublished.

Clinical outcomes related to custom-based components mainly concerned the aMace® acetabular revision system (Materialise, Leuven, Belgium). Colen et al. [103] reported of the successful application in the short-term of this custom design used in six patients with severe acetabular defects (mean age 69.2 years). Similarly, Citak et al. [21] and Baauw et al. [105] described the clinical outcomes of the aMace® solution in nine and twelve patients, respectively. Encouraging results were showed, although in the first study one case needed a further operation due to implant failure at 13 months. The complexity of the patient condition must be taken into account, considering that the average number of revisions that patients underwent before being implanted with the custom components were reported to be five and two, respectively. Baauw et al. [104] reported also of the difficulty to achieve a correct position of the patient-specific implant comparing pre-surgery plan and post-surgery clinical outcome with CT scans; seven out of sixteen cases showed malposition in at least one of inclination, anteversion, rotation or position of the centre of rotation. Other good clinical results on the use of this implant were reported by Wong et al. [78] and Goriainov et al. [106], with one and eleven patients, respectively. The challenges in planning and implementation of the operation were highlighted, but satisfactory outcomes were reached in treating patients with disability or severe bone loss. Myncke et al. [107] reported about the experiences in using the aMace® technology in Belgium between September 2009 and November 2014. The overall results suggested that 3D printed custom implant can represent an acceptable solution for complex acetabular revisions, despite the cumbersome and time-consuming planning procedure. A high rate of complications (36%, 8/22 cases) were also reported, concerning mainly dislocations; this was attributed to the poor quality of the soft tissue in multioperated patients. Lastly, Angelini et al. [108] reported satisfactory early clinical outcomes (mean follow-up 13.7 months) using both Promade and C-Fit 3D® in six patients with severe acetabular defects. In general, likewise off-the-shelf implants, only clinical outcomes in the short-term have been reported for custom-based components. Moreover, considering the high costs, these implants have been used in a limited number of cases where the acetabular defects could not be corrected with any other solution. Longer follow-ups and bigger numbers are required to better understand the performance of AM patient-specific implants.

5. Conclusions

In this review of 3D printed acetabular implants, we compared and contrasted 3D printing with conventional manufacturing. We defined the clinical and engineering rationale of 3D printed acetabular cups, summarized the key variables involved in the manufacturing process, and created a classification of the applications. Whilst early clinical outcomes related to 3D printed cups have been promising, in-depth robust investigations are needed.

3D printing has already revolutionized the way customized orthopaedic implants are produced and the same may happen for off-the-shelf implants. Complex porous structures to achieve enhanced fixation with bone and the printing of personalized shapes are the two main advantages of 3D printing over conventional manufacture for customized implants. Enhanced porous structures and design flexibility is the main rationale for the use of off-the-shelf implants.

However, the importance of standardization of all the steps involved in the production process before implantation in the human body is clear. Both the International Organization for Standardization (ISO) and the American Society for Testing and Material (ASTM) have released guidelines related to terminology, design, process, materials and test methods [25,31,110], but significant gaps remain with respect to the interplay of various parameters in the properties of the final part and how to investigate the potential presence of defects.

In-depth analysis of the properties of these implants is highly suggested in order to avoid unexpected wide-scale failures as occurred with metal-on-metal implants [10,11]. A better understanding of the link between the microstructure, processing and properties of 3D printed components is necessary, considering the complex thermal cycles and the distinctive layer-over-layer building experienced with this technology, which are issues uniquely associated with the use of 3D printer technology [6].

Although it is still unclear if this new manufacturing technology is completely suitable for orthopaedic implants, further analysis of both pristine and retrieved components, together with long-term clinical outcomes and surveillance of new implants placed on the market, will help the transition to 3D printing to be managed safely.

Author Contributions: Conceptualization, L.D., H.H., A.D.L., J.H., A.H.; writing—original draft preparation, L.D.; writing—review and editing, H.H., A.D.L., J.H., A.H.

Acknowledgments: This review was supported by The Maurice Hatter Foundation, the RNOH Charity, the Rosetrees Trust and the Stoneygate Trust and the National Institute for Health Research University College London Hospitals Biomedical Research Centre.

Conflicts of Interest: The authors declare no conflict of interest.

References

1. Abdel Jaber, S.; Affatato, S. An overview of in vitro mechanical and structural characterization of hip prosthesis components. *Biomater. Clin. Pract. Adv. Clin. Res. Med. Devices* **2017**, 585–599. [CrossRef]
2. Bozic, K.J.; Kurtz, S.M.; Lau, E.; Ong, K.; Vail, T.P.; Berry, D.J. The Epidemiology of Revision Total Hip Arthroplasty in the United States. *J. Bone Jt. Surg. Am. Vol.* **2009**, *91*, 128–133. [CrossRef] [PubMed]

3. Patel, A.; Pavlou, G.; Mújica-Mota, R.E.; Toms, A.D. The epidemiology of revision total knee and hip arthroplasty in England and Wales: A comparative analysis with projections for the United States. A study using the national joint registry dataset. *Bone Jt. J.* **2015**, *97-B*, 1076–1081. [CrossRef] [PubMed]
4. Banerjee, S.; Kulesha, G.; Kester, M.; Mont, M.A. Emerging technologies in arthroplasty: Additive manufacturing. *J. Knee Surg.* **2014**, *27*, 185–191. [CrossRef] [PubMed]
5. Mumith, A.; Thomas, M.; Shah, Z.; Coathup, M.; Blunn, G. Additive manufacturing. *Bone Jt. J.* **2018**, *100-B*, 455–460. [CrossRef] [PubMed]
6. Frazier, W.E. Metal additive manufacturing: A review. *J. Mater. Eng. Perform.* **2014**, *23*, 1917–1928. [CrossRef]
7. Murr, L.E.; Gaytan, S.M.; Martinez, E.; Medina, F.; Wicker, R.B. Next generation orthopaedic implants by additive manufacturing using electron beam melting. *Int. J. Biomater.* **2012**, *2012*, 1–14. [CrossRef] [PubMed]
8. Morrison, R.J.; Kashlan, K.N.; Flanangan, C.L.; Wright, J.K.; Green, G.E.; Hollister, S.J.; Weatherwax, K.J. Regulatory considerations in the design and manufacturing of implantable 3D-printed medical devices. *Clin. Transl. Sci.* **2015**, *8*, 594–600. [CrossRef]
9. Orthopaedic Data Evaluation Panel (ODEP). Available online: http://www.odep.org.uk/products.aspx (accessed on 1 April 2019).
10. Hothi, H.S.; Berber, R.; Panagiotopoulos, A.C.; Whittaker, R.K.; Rhead, C.; Skinner, J.A.; Hart, A.J. Clinical significance of corrosion of cemented femoral stems in metal-on-metal hips: A retrieval study. *Int. Orthop.* **2016**, *40*, 2247–2254. [CrossRef]
11. Hothi, H.S.; Berber, R.; Whittaker, R.K.; Blunn, G.W.; Skinner, J.A.; Hart, A.J. The relationship between cobalt/chromium ratios and the high prevalence of head-stem junction corrosion in metal-on-metal total hip arthroplasty. *J. Arthroplasty* **2016**, *31*, 1123–1127. [CrossRef]
12. National Joint Registry for England, Wales, Northern Ireland and the Isle of Man: 15th Annual Report. 2018. Available online: www.njrreports.org.uk (accessed on 1 April 2019).
13. Murr, L.E.; Gaytan, S.M.; Medina, F.; Lopez, H.; Martinez, E.; MacHado, B.I.; Hernandez, D.H.; Martinez, L.; Lopez, M.I.; Wicker, R.B.; et al. Next-generation biomedical implants using additive manufacturing of complex cellular and functional mesh arrays. *Philos. Trans. R. Soc. A Math. Phys. Eng. Sci.* **2010**, *368*, 1999–2032. [CrossRef] [PubMed]
14. Dennis, D.A. Management of massive acetabular defects in revision total hip arthroplasty. *J. Arthropl.* **2003**, *18*, 121–125. [CrossRef] [PubMed]
15. Jain, S.; Grogan, R.J.; Giannoudis, P.V. Options for managing severe acetabular bone loss in revision hip arthroplasty. A systematic review. *Hip Int.* **2014**, *24*, 109–122. [CrossRef] [PubMed]
16. Moore, K.D.; McClenny, M.D.; Wills, W.B. Custom Triflange acetabular components for large acetabular defects: Minimum 10-year follow-up. *Orthopedics* **2018**, *16*, 1–5. [CrossRef] [PubMed]
17. Hart, A.J.; Hart, A.; Panagiotopoulou, V.; Henckel, J. *Orthopaedic Products News*, 2017; 178, 22–26.
18. Sheth, N.P.; Nelson, C.L.; Springer, B.D.; Fehring, T.K.; Paprosky, W.G. Acetabular bone loss in revision total hip arthroplasty: Evaluation and management. *J. Am. Acad. Orthop. Surg.* **2013**, *21*, 128–139. [CrossRef] [PubMed]
19. Learmonth, I.D.; Young, C.; Rorabeck, C. The operation of the century: total hip replacement. *Lancet* **2007**, *370*, 1508–1519. [CrossRef]
20. Paprosky, W.G.; Perona, P.G.; Lawrence, J.M. Acetabular defect classification and surgical reconstruction in revision arthroplasty. A 6-year follow-up evaluation. *J. Arthropl.* **1994**, *9*, 33–44. [CrossRef]
21. Citak, M.; Kochsiek, L.; Gehrke, T.; Haasper, C.; Suero, E.M.; Mau, H. Preliminary results of a 3D-printed acetabular component in the management of extensive defects. *Hip Int.* **2017**, *28*, 266–271. [CrossRef]
22. Frank, R.M.; Fabi, D.; Levine, B.R. Modern porous coatings in orthopaedic applications. In *Thin Films and Coatings in Biology*; Springer: Cham, Switzerland, 2013; pp. 69–103. ISBN 978-94-007-2591-1.
23. Burroughs, B.R.; Hallstrom, B.; Golladay, G.J.; Hoeffel, D.; Harris, W.H. Range of motion and stability in total hip arthroplasty with 28-, 32-, 38-, and 44-mm femoral head sizes: An in vitro study. *J. Arthropl.* **2005**, *20*, 11–19. [CrossRef]
24. Berry, D.J.; Von Knoch, M.; Schleck, C.D.; Harmsen, W.S.; Knoch, M. Von dislocation after primary total hip arthroplasty effect of femoral head diameter and operative approach on risk of effect of femoral head diameter and operative approach on risk of dislocation after primary total hip arthroplasty. *J. Bone Jt. Surg.* **2005**, 2456–2463. [CrossRef]

25. ASTM 52900:2015(E): American Society for Testing and Materials (ASTM). *Standard Terminology for Additive Manufacturing—General Principles—Terminology*; ASTM: West Conshohocken, PA, USA, 2015.
26. Zhai, Y.; Lados, D.A.; Lagoy, J.L. Additive manufacturing: Making imagination the major limitation. *Jom* **2014**, *66*, 808–816. [CrossRef]
27. McTighe, T.; Brazil, D.; Bruce, W. Metallic Alloys in Total Hip Arthroplasty. In *The Hip: Preservation, Replacement and Revision*; Parvizi, J., Goyal, N., Cashman, J., Eds.; Data Trace Publishing Company: Brooklandville, MD, USA, 2015; pp. 1–12. ISBN 9781574001495.
28. Long, M.; Rack, H.J. Titanium alloys in total joint replacement—A materials science perspective. *Biomaterials* **1998**, *19*, 1621–1639. [CrossRef]
29. International Organization for Standardization. *Implants for Surgery—Metallic Materials-Part 3: Wrought Titanium 6-Aluminium 4-Vanadium Alloy*; ISO 5832-3; International Organization for Standardization: Geneva, Switzerland, 2016.
30. ASTM F1472-14: American Society for Testing and Materials (ASTM). *Standard Specification for Wrought Titanium-6Aluminum-4Vanadium Alloy for Surgical Implant Applications*; ASTM: West Conshohocken, PA, USA, 2014.
31. ASTM F2924-14: American Society for Testing and Materials (ASTM). *Standard Specification for Additive Manufacturing Titanium-6 Aluminum-4 Vanadium with Powder Bed Fusion*; ASTM: West Conshohocken, PA, USA, 2014.
32. Murr, L.E.; Quinones, S.A.; Gaytan, S.M.; Lopez, M.I.; Rodela, A.; Martinez, E.Y.; Hernandez, D.H.; Martinez, E.; Medina, F.; Wicker, R.B. Microstructure and mechanical behavior of Ti-6Al-4V produced by rapid-layer manufacturing, for biomedical applications. *J. Mech. Behav. Biomed. Mater.* **2009**, *2*, 20–32. [CrossRef] [PubMed]
33. Ryan, G.; Pandit, A.; Apatsidis, D.P. Fabrication methods of porous metals for use in orthopaedic applications. *Biomaterials* **2006**, *27*, 2651–2670. [CrossRef]
34. Wang, X.; Xu, S.; Zhou, S.; Xu, W.; Leary, M.; Choong, P.; Qian, M.; Brandt, M.; Xie, Y.M. Topological design and additive manufacturing of porous metals for bone scaffolds and orthopaedic implants: A review. *Biomaterials* **2016**, *83*, 127–141. [CrossRef] [PubMed]
35. Muth, J.; Poggie, M.; Kulesha, G.; Meneghini, R.M. Novel highly porous metal technology in artificial hip and knee replacement: Processing methodologies and clinical applications. *Jom* **2013**, *65*, 318–325. [CrossRef]
36. Vaithilingam, J.; Prina, E.; Goodridge, R.D.; Hague, R.J.M.; Edmondson, S.; Rose, F.R.A.J.; Christie, S.D.R. Surface chemistry of Ti6Al4V components fabricated using selective laser melting for biomedical applications. *Mater. Sci. Eng. C* **2016**, *67*, 294–303. [CrossRef] [PubMed]
37. Sing, S.L.; An, J.; Yeong, W.Y.; Wiria, F.E. Laser and electron-beam powder-bed additive manufacturing of metallic implants: A review on processes, materials and designs. *J. Orthop. Res.* **2016**, *34*, 369–385. [CrossRef] [PubMed]
38. Heinl, P.; Rottmair, A.; Körner, C.; Singer, R.F. Cellular titanium by selective electron beam melting. *Adv. Eng. Mater.* **2007**, *9*, 360–364. [CrossRef]
39. Murr, L.E. Additive manufacturing of biomedical devices: An overview. *Mater. Technol.* **2018**, *33*, 57–70. [CrossRef]
40. Parthasarathy, J.; Starly, B.; Raman, S.; Christensen, A. Mechanical evaluation of porous titanium (Ti6Al4V) structures with electron beam melting (EBM). *J. Mech. Behav. Biomed. Mater.* **2010**, *3*, 249–259. [CrossRef] [PubMed]
41. Heinl, P.; Müller, L.; Körner, C.; Singer, R.F.; Müller, F.A. Cellular Ti-6Al-4V structures with interconnected macro porosity for bone implants fabricated by selective electron beam melting. *Acta Biomater.* **2008**, *4*, 1536–1544. [CrossRef] [PubMed]
42. Arcam EBM a GE Additive Company. Available online: www.arcam.com (accessed on 1 April 2019).
43. Sames, W.J.; List, F.A.; Pannala, S.; Dehoff, R.R.; Babu, S.S. The metallurgy and processing science of metal additive manufacturing. *Int. Mater. Rev.* **2016**, *61*, 315–360. [CrossRef]
44. Gibson, I.; Rosen, D.; Stucker, B. Powder bed fusion processes. In *Additive Manufacturing Technologies*; Springer: Cham, Switzerland, 2015; pp. 107–145. ISBN 978-1-4939-2112-6.
45. DebRoy, T.; Wei, H.L.; Zuback, J.S.; Mukherjee, T.; Elmer, J.W.; Milewski, J.O.; Beese, A.M.; Wilson-Heid, A.; De, A.; Zhang, W. Additive manufacturing of metallic components—Process, structure and properties. *Prog. Mater. Sci.* **2018**, *92*, 112–224. [CrossRef]

46. O'Regan, P.; Prickett, P.; Setchi, R.; Hankins, G.; Jones, N. Metal based additive layer manufacturing: Variations, correlations and process control. *Procedia Comput. Sci.* **2016**, *96*, 216–224. [CrossRef]
47. Leary, M. Design of titanium implants for additive manufacturing. In *Titanium in Medical and Dental Applications*; Elsevier Inc.: Amsterdam, The Netherlands, 2018; pp. 203–224. ISBN 9780128124567.
48. Tang, H.P.; Qian, M.; Liu, N.; Zhang, X.Z.; Yang, G.Y.; Wang, J. Effect of powder reuse times on additive manufacturing of ti-6al-4v by selective electron beam melting. *Jom* **2015**, *67*, 555–563. [CrossRef]
49. Donachie, M. *Titanium—A Technical Guide*, 2nd ed.; ASM International: Russell Township, OH, USA, 2000.
50. Huotilainen, E.; Jaanimets, R.; Valášek, J.; Marcián, P.; Salmi, M.; Tuomi, J.; Mäkitie, A.; Wolff, J. Inaccuracies in additive manufactured medical skull models caused by the DICOM to STL conversion process. *J. Cranio-Maxillofacial Surg.* **2014**, *42*, 259–265. [CrossRef] [PubMed]
51. Lieneke, T.; Adam, G.A.O.; Leuders, S.; Knoop, F.; Josupeit, S.; Delfs, P.; Funke, N.; Zimmer, D. Systematical determination of tolerances for additive manufacturing by measuring linear dimensions. In Proceedings of the Annual International Solid Freeform Fabrication Symposium, Austin, TX, USA, 8–12 May 2015; pp. 371–384.
52. Murr, L.E.; Gaytan, S.M.; Ramirez, D.A.; Martinez, E.; Hernandez, J.; Amato, K.N.; Shindo, P.W.; Medina, F.R.; Wicker, R.B. Metal fabrication by additive manufacturing using laser and electron beam melting technologies. *J. Mater. Sci. Technol.* **2012**, *28*, 1–14. [CrossRef]
53. Marin, E.; Fusi, S.; Pressacco, M.; Paussa, L.; Fedrizzi, L. Characterization of cellular solids in Ti6Al4V for orthopaedic implant applications: Trabecular titanium. *J. Mech. Behav. Biomed. Mater.* **2010**, *3*, 373–381. [CrossRef]
54. Regis, M.; Marin, E.; Fusi, S.; Pressacco, M.; Fedrizzi, L. Preparation and characterization of newly developed trabecular structures in titanium alloy to optimize osteointegration. *Int. J. Med. Health Pharm. Biomed. Eng.* **2014**, *8*, 279–284.
55. Zhao, X.; Li, S.; Zhang, M.; Liu, Y.; Sercombe, T.B.; Wang, S.; Hao, Y.; Yang, R.; Murr, L.E. Comparison of the microstructures and mechanical properties of Ti-6Al-4V fabricated by selective laser melting and electron beam melting. *Mater. Des.* **2016**, *95*, 21–31. [CrossRef]
56. Hrabe, N.; Quinn, T. Effects of processing on microstructure and mechanical properties of a titanium alloy (Ti-6Al-4V) fabricated using electron beam melting (EBM), Part 2: Energy input, orientation, and location. *Mater. Sci. Eng. A* **2013**, *573*, 271–277. [CrossRef]
57. Galarraga, H.; Lados, D.A.; Dehoff, R.R.; Kirka, M.M.; Nandwana, P. Effects of the microstructure and porosity on properties of Ti-6Al-4V ELI alloy fabricated by electron beam melting (EBM). *Addit. Manuf.* **2016**, *10*, 47–57. [CrossRef]
58. Concept Laser a GE Additive Company. Available online: www.conceptlaserinc.com (accessed on 1 April 2019).
59. Murr, L.E.; Martinez, E.; Amato, K.N.; Gaytan, S.M.; Hernandez, J.; Ramirez, D.A.; Shindo, P.W.; Medina, F.; Wicker, R.B. Fabrication of metal and alloy components by additive manufacturing: Examples of 3D materials science. *J. Mater. Res. Technol.* **2012**, *1*, 42–54. [CrossRef]
60. Wang, P.; Sin, W.J.; Nai, M.L.S.; Wei, J. Effects of processing parameters on surface roughness of additive manufactured Ti-6Al-4V via electron beam melting. *Materials* **2017**, *10*, 1121. [CrossRef]
61. Gong, H.; Rafi, K.; Gu, H.; Starr, T.; Stucker, B. Analysis of defect generation in Ti-6Al-4V parts made using powder bed fusion additive manufacturing processes. *Addit. Manuf.* **2014**, *1*, 87–98. [CrossRef]
62. King, W.E.; Barth, H.D.; Castillo, V.M.; Gallegos, G.F.; Gibbs, J.W.; Hahn, D.E.; Kamath, C.; Rubenchik, A.M. Observation of keyhole-mode laser melting in laser powder-bed fusion additive manufacturing. *J. Mater. Process. Technol.* **2014**, *214*, 2915–2925. [CrossRef]
63. Kasperovich, G.; Haubrich, J.; Gussone, J.; Requena, G. Correlation between porosity and processing parameters in TiAl6V4 produced by selective laser melting. *Mater. Des.* **2016**, *105*, 160–170. [CrossRef]
64. Mukherjee, T.; Zuback, J.S.; De, A.; DebRoy, T. Printability of alloys for additive manufacturing. *Sci. Rep.* **2016**, *6*, 19717. [CrossRef]
65. Wong, K.-C.; Scheinemann, P. Additive manufactured metallic implants for orthopaedic applications. *Sci. China Mater.* **2018**, *61*, 440–454. [CrossRef]
66. Culmone, C.; Smit, G.; Breedveld, P. Additive manufacturing of medical instruments: A state-of-the-art review. *Addit. Manuf.* **2019**, *27*, 461–473. [CrossRef]

67. Du Plessis, A.; le Roux, S.G.; Booysen, G.; Els, J. Quality control of a laser additive manufactured medical implant by X-Ray tomography. *3D Print. Addit. Manuf.* **2016**, *3*, 175–182. [CrossRef]
68. Martelli, N.; Serrano, C.; Van Den Brink, H.; Pineau, J.; Prognon, P.; Borget, I.; El Batti, S. Advantages and disadvantages of 3-dimensional printing in surgery: A systematic review. *Surgery* **2016**, *159*, 1485–1500. [CrossRef] [PubMed]
69. Di Prima, M.; Coburn, J.; Hwang, D.; Kelly, J.; Khairuzzaman, A.; Ricles, L. Additively manufactured medical products—The FDA perspective. *3D Print. Med.* **2015**, *2*, 1–6. [CrossRef] [PubMed]
70. Adler Ortho. Available online: www.adlerortho.com (accessed on 1 April 2019).
71. OsseoTi TM Porous Metal for Enhanced Bone Integration an Animal Study. Available online: www.zimmerbiomet.com (accessed on 1 April 2019).
72. Stryker. Available online: www.stryker.com (accessed on 1 April 2019).
73. Castagnini, F.; Bordini, B.; Stea, S.; Calderoni, P.P.; Masetti, C.; Busanelli, L. Highly porous titanium cup in cementless total hip arthroplasty: Registry results at eight years. *Int. Orthop.* **2018**, 1–7. [CrossRef] [PubMed]
74. Corin Connected Orthopaedic Insight. Available online: www.coringroup.com (accessed on 1 April 2019).
75. Implantcast. Available online: www.implantcast.de (accessed on 1 April 2019).
76. Lima Corporate Orthopaedic Emotion. Available online: www.limacorporate.com (accessed on 1 April 2019).
77. Mobelife. *aMace Acetabular Revision System*, 2014.
78. Wong, K.C.; Kumta, S.M.; Gee, N.V.L.; Demol, J. One-step reconstruction with a 3D-printed, biomechanically evaluated custom implant after complex pelvic tumor resection. *Comput. Aided Surg.* **2015**, *20*, 14–23. [CrossRef] [PubMed]
79. Medacta International. Available online: www.medacta.com (accessed on 1 April 2019).
80. Conceloc Advanced Porous Titanium Material Specification. Available online: www.medacta.com (accessed on 1 April 2019).
81. Zhang, Z.; Jones, E.; Lee, P.D.; Jones, J.R.; Kim, T.B.; Yue, S. Additive manufactured porous titanium structures: Through-process quantification of pore and strut networks. *J. Mater. Process. Technol.* **2014**, *214*, 2706–2715. [CrossRef]
82. Jasty, M.; Bragdon, C.; Burke, D.; O'Connor, D.; Lowenstein, J.; Harris, W.H. In vivo skeletal responses to porous-surfaced implants subjected to small induced motions. *J. Bone Jt. Surg.Ser. A* **1997**, *79*, 707–714. [CrossRef] [PubMed]
83. Karageorgiou, V.; Kaplan, D. Porosity of 3D biomaterial scaffolds and osteogenesis. *Biomaterials* **2005**, *26*, 5474–5491. [CrossRef] [PubMed]
84. Koschwanez, H.E.; Reichert, W.M. Textured and porous materials. In *Biomaterials Science: An Introduction to Materials*, 3rd ed.; Elsevier: Amsterdam, The Netherlands, 2013; pp. 321–331. ISBN 9780123746269.
85. Pałka, K.; Pokrowiecki, R. Porous titanium implants: A review. *Adv. Eng. Mater.* **2018**, *20*, 1–18. [CrossRef]
86. Marin, E.; Pressacco, M.; Fusi, S.; Lanzutti, A.; Turchet, S.; Fedrizzi, L. Characterization of grade 2 commercially pure Trabecular Titanium structures. *Mater. Sci. Eng. C* **2013**, *33*, 2648–2656. [CrossRef] [PubMed]
87. Regis, M.; Marin, E.; Fedrizzi, L.; Pressacco, M. Additive manufacturing of Trabecular Titanium orthopedic implants. *MRS Bull.* **2015**, *40*, 137–144. [CrossRef]
88. Davignon, R.; Rajaravivarma, R.; Yanoso-Scholl, L.; Gopalakrishnan, A.; Abitante, P. Comparison of seating and initial stability of an additive manufactured porous acetabular shell to a clinically successful porous acetabular shell. In Proceedings of the 30th Annual Congress of the International Society for Technology in Arthroplasty, Seoul, South Korea, 20–23 September 2017; pp. 146–148.
89. Kourra, N.; Warnett, J.M.; Attridge, A.; Dibling, G.; McLoughlin, J.; Muirhead-Allwood, S.; King, R.; Williams, M.A. Computed tomography metrological examination of additive manufactured acetabular hip prosthesis cups. *Addit. Manuf.* **2018**, *22*, 146–152. [CrossRef]
90. Jahnke, A.; Bott, C.C.; Fonseca Ulloa, C.A.; Jahnke, G.W.; Rickert, M.; Ishaque, B.A.; Ahmed, G.A. In vitro examination of the primary stability of three press-fit acetabular cups under consideration of two different bearing couples. *Med. Eng. Phys.* **2019**. [CrossRef] [PubMed]
91. Le Cann, S.; Galland, A.; Rosa, B.; Le Corroller, T.; Pithioux, M.; Argenson, J.N.; Chabrand, P.; Parratte, S. Does surface roughness influence the primary stability of acetabular cups? A numerical and experimental biomechanical evaluation. *Med. Eng. Phys.* **2014**, *36*, 1185–1190. [CrossRef] [PubMed]

92. Taniguchi, N.; Fujibayashi, S.; Takemoto, M.; Sasaki, K.; Otsuki, B.; Nakamura, T.; Matsushita, T.; Kokubo, T.; Matsuda, S. Effect of pore size on bone ingrowth into porous titanium implants fabricated by additive manufacturing: An in vivo experiment. *Mater. Sci. Eng. C* **2016**, *59*, 690–701. [CrossRef] [PubMed]
93. Palmquist, A.; Shah, F.A.; Emanuelsson, L.; Omar, O.; Suska, F. A technique for evaluating bone ingrowth into 3D printed, porous Ti6Al4V implants accurately using X-ray micro-computed tomography and histomorphometry. *Micron* **2017**, *94*, 1–8. [CrossRef] [PubMed]
94. Palmquist, A.; Emanuelsson, L.; Thomsen, P.; Palmquist, A.; Snis, A.; Emanuelsson, L.; Thomsen, P.; Snis, A.; Browne, M. Long-term bioçompatibility and osseointegration of electron beam melted, free-form–fabricated solid and porous titanium alloy: Experimental studies in sheep. *J. Biomater. Appl.* **2013**, *27*, 1003–1016. [CrossRef] [PubMed]
95. Delta TT Acetabular System (K112898). Available online: www.fda.gov (accessed on 1 April 2019).
96. Mpact®3D MetalTM Implants and Augments 3D MetalTM (K171966). Available online: www.fda.gov (accessed on 1 April 2019).
97. Trident®II Tritanium®Acetabular Shell (K161569). Available online: www.fda.gov (accessed on 1 April 2019).
98. Steno, B.; Kokavec, M.; Necas, L. Acetabular revision arthroplasty using trabecular titanium implants. *Int. Orthop.* **2015**, *39*, 389–395. [CrossRef]
99. Perticarini, L.; Zanon, G.; Rossi, S.M.P.; Benazzo, F.M. Clinical and radiographic outcomes of a trabecular titanium^{TM} acetabular component in hip arthroplasty: Results at minimum 5 years follow-up. *BMC Musculoskelet. Disord.* **2015**, *16*, 375. [CrossRef]
100. Massari, L.; Bistolfi, A.; Grillo, P.P.; Borré, A.; Gigliofiorito, G.; Pari, C.; Francescotto, A.; Tosco, P.; Deledda, D.; Ravera, L.; et al. Periacetabular bone densitometry after total hip arthroplasty with highly porous titanium cups: A 2-year follow-up prospective study. *HIP Int.* **2017**, *27*, 551–557. [CrossRef] [PubMed]
101. Gallart, X.; Fernández-Valencia, J.A.; Riba, J.; Bori, G.; García, S.; Tornero, E.; Combalía, A. Trabecular Titanium^{TM} cups and augments in revision total hip arthroplasty: Clinical results, radiology and survival outcomes. *HIP Int.* **2016**, *26*, 486–491. [CrossRef] [PubMed]
102. De Meo, F.; Cacciola, G.; Bellotti, V.; Bruschetta, A.; Cavaliere, P. Trabecular Titanium acetabular cups in hip revision surgery: Mid-term clinical and radiological outcomes. *HIP Int.* **2018**, *28*, 61–65. [CrossRef] [PubMed]
103. Colen, S.; Harake, R.; De Haan, J.; Mulier, M. A modified custom-made triflanged acetabular reconstruction ring (MCTARR) for revision hip arthroplasty with severe acetabular defects. *Acta Orthop. Belg.* **2013**, *79*, 71–75. [PubMed]
104. Baauw, M.; Van Hellemondt, G.G.; Van Hooff, M.L.; Spruit, M. The accuracy of positioning of a custom-made implant within a large acetabular defect at revision arthroplasty of the hip. *Bone Jt. J.* **2015**, *97-B*, 780–785. [CrossRef] [PubMed]
105. Baauw, M.; van Hellemondt, G.G.; Spruit, M. A Custom-made Acetabular Implant for Paprosky Type 3 Defects. *Orthopedics* **2017**, *40*, e195–e198. [CrossRef] [PubMed]
106. Goriainov, V.; McEwan, J.K.; Oreffo, R.O.; Dunlop, D.G. Application of 3D-printed patient-specific skeletal implants augmented with autologous skeletal stem cells. *Regen. Med.* **2018**, *13*, 283–294. [CrossRef]
107. Myncke, I.; Van Schaik, D.; Scheerlinck, T. Custom-made triflanged acetabular components in the treatment of major acetabular defects. Short-term results and clinical experience. *Acta Orthop. Belg.* **2017**, *83*, 341–350.
108. Angelini, A.; Trovarelli, G.; Berizzi, A.; Pala, E.; Breda, A.; Ruggieri, P. Three-dimension-printed custom-made prosthetic reconstructions: From revision surgery to oncologic reconstructions. *Int. Orthop.* **2018**, *43*, 123–132. [CrossRef]
109. Harris, W. Traumatic Arthritis of the hip after dislocation and acetabular fractures: treatment by mold arthroplasty. *J. Bone Jt. Surg.* **1969**, *51*, 737–755. [CrossRef]
110. ISO 17296-3. *Additive manufacturing—General principles Part 3: Main Characteristics and Corresponding Test Methods*; International Organization for Standardization: Geneva, Switzerland, 2014.

metals

MDPI

Review

On the Corrosion Behaviour of Low Modulus Titanium Alloys for Medical Implant Applications: A Review

Pooria Afzali [1], Reza Ghomashchi [2] and Reza H. Oskouei [3,*]

1 Faculty of Materials and Metallurgical Engineering, Semnan University, Semnan 35131-19111, Iran
2 School of Mechanical Engineering, University of Adelaide, Adelaide, SA 5000, Australia
3 College of Science and Engineering, Flinders University, Clovelly Park, SA 5042, Australia
* Correspondence: reza.oskouei@flinders.edu.au; Tel.: +61-8-8201-2782

Received: 2 July 2019; Accepted: 6 August 2019; Published: 9 August 2019

check for updates

Abstract: The corrosion behaviour of new generation titanium alloys (β-type with low modulus) for medical implant applications is of paramount importance due to their possible detrimental effects in the human body such as release of toxic metal ions and corrosion products. In spite of remarkable advances in improving the mechanical properties and reducing the elastic modulus, limited studies have been done on the electrochemical corrosion behaviour of various types of low modulus titanium alloys including the effect of different beta-stabilizer alloying elements. This development should aim for a good balance between mechanical properties, design features, metallurgical aspects and, importantly, corrosion resistance. In this article, we review several significant factors that can influence the corrosion resistance of new-generation titanium alloys such as fabrication process, body electrolyte properties, mechanical treatments, alloying composition, surface passive layer, and constituent phases. The essential factors and their critical features are discussed. The impact of various amounts of α and β phases in the microstructure, their interactions, and their dissolution rates on the surface passive layer and bulk corrosion behaviour are reviewed and discussed in detail. In addition, the importance of different corrosion types for various medical implant applications is addressed in order to specify the significance of every corrosion phenomenon in medical implants.

Keywords: low modulus titanium alloys; corrosion; medical implants; young's modulus

---

1. Introduction

The need for biocompatible materials with a good combination of mechanical and electrochemical properties has been always of great significance for manufacturing medical implants, particularly load-bearing orthopaedic implants used for joint arthroplasty. Producing implants that can enhance the longevity in the body is imperative which, in turn, can reduce the possibility of implant failure; and hence, risky and expensive revision surgeries. Titanium and its alloys have long been used for biomedical implant applications as they offer a favourable set of properties including high specific strength, good biocompatibility and high corrosion resistance [1–3]. In terms of osteogenesis, titanium and its alloys have been reported to possess the greatest biocompatibility in comparison with the other biometals [4]. Nowadays, Ti-6Al-4V alloys are extensively used for the manufacture of load-bearing orthopaedic implants such as hip stems that are to live with the surrounding hard and soft tissues in the physiological media of the body [5,6].

Despite all the aforementioned advantages of Ti-6Al-4V alloy, research has identified some concerns around the toxicity of aluminium and vanadium present in this alloy. Neurological disorders, allergic reactions, Alzheimer's disease and cytotoxicity are of some drawbacks associated with the release of Al and V into the body [7–11]. Also, from the mechanical behaviour perspective, the Young's modulus of Ti-6Al-4V is approximately 110 GPa [12] that is significantly higher than the stiffness of the bone (typically 10–30 GPa) [5]. It is known that if there is a substantial discrepancy between the stiffness of the implant and the adjacent bone, significant stress shielding can occur under the mechanical loads of daily activities which can result in adverse effects such as bone loss, implant loosening, and peri-prosthetic fracture [11,13–17].

In order to address the issues associated with the most commonly used alloy of Ti-6Al-4V, there have been many efforts to develop a new class of titanium alloys (known as β-type titanium alloys) with reduced levels of stiffness. Any alteration in the chemical composition of Ti alloys may lead to stabilization of a certain phase and crystal structure: the high temperature Ti has a body-centered cubic (BCC) crystal structure, β-phase, while the low temperature phase (α) displays a hexagonal close-packed (HCP) structure (e.g., CP Ti) and the combination of the two phases (α + β) (e.g., Ti-6Al-4V) [18]. The β-type titanium alloys containing beta-stabilizing elements (e.g., Nb, Ta and Zr) exhibit many advantages such as lower Young's moduli that are closer to that of the human bone (which can mitigate bone loss and implant loosening due to stress shielding), non-allergic and non-toxic elements such as Nb, Ta and Zr, excellent corrosion resistance due to the formation of more stable oxide layers and good biocompatibility [19,20]. Body fluid present in the immediate vicinity of implants contains organic and inorganic solutions of cations Mg^{2+}, Ca^{2+}, Na^{+}, K^{+}, and anions $SO_4{}^{2-}$, Cl^{-}, $HCO_3{}^{-}$, $H_2PO_4{}^{-}$. Also, from the corrosion science perspective, due to the presence of such ions, the pH of body fluid may fluctuate from acidic to alkaline; and this can cause the open circuit potential (E_{oc}) to have different values in various parts of the implant surface. Hence, the adequate driving force of localized corrosion may increase which can speed up the degradation of the implant material [21].

Titanium alloys spontaneously form a passive oxide layer (mainly TiO_2) that, to some extent, can protect these alloys against corrosion because of its thermodynamic stability, chemical inertia, and low solubility in the body fluids [22]. However, severe corrosion can occur when this passive oxide layer is mechanically disrupted. In such situations, aggressive Cl^{-} ions may attack the implant material surface. For instance, fretting wear is known to occur at the interface of modular junctions (for example, head-neck taper junction of hip implants) due to relative micro-motions [23–26]. The fretting wear disrupts the passive oxide layer causing corrosion to occur in the alloy in vivo. Moreover, it is important to mention that metallic ions released due to the corrosion process can cause adverse reactions in the body resulting in implant revision surgeries due to the release of toxic ions in the body [23,27–29].

Now that new generation Ti alloys have been already developed with lower Young's moduli (much closer to bone when compared to widely used Ti-6Al-4V) and that they have shown good biocompatibility, it is important to review their corrosion behaviour and electrochemical characteristics. There are various groups of new generation titanium alloys that may offer different levels of corrosion resistance in the body fluid depending on their chemistry (alloying elements and their amounts), and fabrication methods. Therefore, this review aims to provide useful information on the corrosion response of various Ti alloys that are increasingly being developed for the next generation medical implants.

2. In Vitro Electrochemical Corrosion Testing Procedures

The electrochemical corrosion testing procedures used for biometals (including low stiffness titanium alloys) involve several various parameters and testing methods depending on the objectives of the experimental work. Potentiodynamic polarization, electrochemical impedance spectroscopy (EIS), open circuit potential (OCP) and other corrosion testing procedures can be applied in order to assess the corrosion behaviour. In addition, parameters such as time of exposure, corrosion medium or solution (electrolyte), pH and the electrolyte temperature can significantly affect the corrosion behaviour of titanium alloys due to the sensitivity of their passive oxide layer. Moreover, microstructural observations of the surface after anodization along with X-ray diffraction (XRD), energy-dispersive X-ray spectroscopy (EDS) and X-ray photoelectron spectroscopy (XPS) examinations can help scientists and researchers to explore corrosion mechanisms and the formation of passive layers with their chemical compositions.

Owing to their passive oxide layer, titanium-based alloys are resistant to some organic acids; for example, lactic acid and acetic acid. However, they can be corroded by some others such as citric acid [30,31]. Among halide anions, fluoride ions can more strongly cause dissolution of the passive oxide layer in titanium-based alloys [32]. It is known that fluoride ions are aggressive and they may cause pitting and crevice corrosion in dental implants [33]. Many commercial oral rinses include significant amounts of fluoride which can affect the electrochemical behaviour of orthodontic implants made of titanium alloys [23]. In a study by Li et al. [34], the effect of fluoride on the electrochemical parameters of Ti-24Nb-4Zr-8Sn (β-type Ti2448) alloy was examined in artificial saliva. Further increase of F^- from (≤0.1 to 1%) degraded the corrosion resistance of the alloy. Also, all the corrosion resistance parameters extracted from the EIS measurements such as R_s (solution resistance), R_b (resistance of the inner barrier oxide layer) and R_p (resistance of the outer porous oxide layer) decreased and were in good agreement with Tafel diagram. Also, NaF concentration was found to influence the protective behaviour of the passive layer onto Ti-6Al-4V causing an increase in the corrosion current density in salvia solution [35]. Also, Huang [35] reported that the electrochemical properties of Ti-6Al-4V alter not only with the fluoride content, but also with the pH and the immersion time in Ringer's solution. Robin and Carvalho [36] investigated the passivation of CP-Ti, Ti-13Nb-13Zr, Ti-5Nb-13Zr and Ti-20Nb-13Zr alloys for a pH range (2–7.5) in a Ringer's solution with (1000 ppm) and without fluoride anions. The corrosion resistance parameters of the tested alloys reduced with a further increase of F^- and decrease of pH. Also, according to the OCP and Bode diagrams, Ti-13Nb-13Zr showed a better corrosion resistance in comparison with the other tested alloys because of a spontaneous stable oxide layer (TiO_2) formed on its surface. Although the chemical stability of Ti-Nb alloys is high (because of their excellent passive layer behaviour [37]), an increase in pH, in acidification of the Ringer's solution from pH 7.5 to pH 0.5 resulted in negative corrosion potential shifts. The addition of NaCl further exacerbated the corrosion resistance [38–42].

Surface microstructural characterization of β-type Ti alloys after electrochemical tests in various simulated body fluid solutions play an important role in verifying the corrosion test results. These observations can reveal corrosion aspects of the surface such as thickness of the passive layer, corrosion product phases and compounds, pitting and corrosion mechanism. Figure 1 shows scanning electron microscope (SEM) images of a Ti-24Nb-4Zr-8Sn alloy after immersion in 0.9% NaCl solution for various durations (24, 72, and 168 h) [36]. The results showed that the surface of the alloy after 24 h included a passive layer with no corrosion attack which was in agreement with the Bode diagram (Figure 2) so that the modulus of impedance at 0.01 Hz frequency increased after 24 h immersion and the phase angle at intermediate to low frequencies was closer to more negative angles in the Bode-phase diagram [43].

Figure 1. Scanning electron microscope (SEM) images of Ti-24Nb-4Zr-8Sn alloy after immersion in 0.9% NaCl solution at 37 °C for: (**a**) 24 h, (**b**) 72 h, and (**c**) 168 h [43].

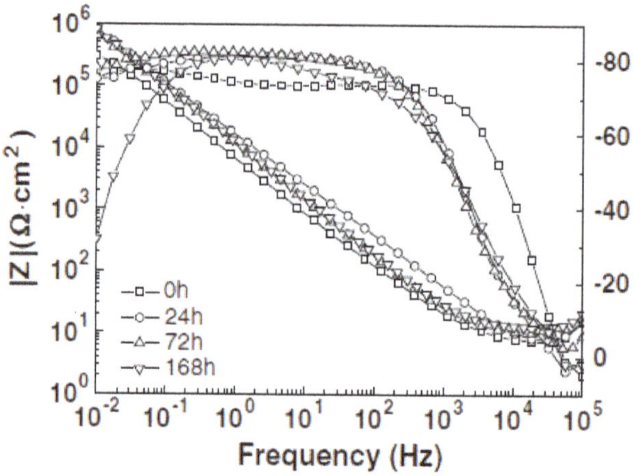

Figure 2. Electrochemical impedance spectroscopy (EIS) spectra of Ti-24Nb-4Zr-8Sn alloy at different immersion hours in 0.9% NaCl solution [43].

SEM images of a biomedical β-type TiNbZrFe alloy after electrochemical testing in 0.9% NaCl and 0.2% NaF solutions showed that the deep corrosion pits in non-SMAT (surface mechanical attrition treated) samples were distributed densely on the surface which was verified with more fluctuations of the anodic branch of the Tafel diagram and the higher corrosion current density. However, the SMAT sample was covered completely with the passive oxide layer [44]. Atapour et al. [40] used secondary electron SEM micrographs of Ti-13Mo-7Zr-3Fe (as-received α + β) and (metastable β) samples after 50 h immersion in 5 M HCl at 37 °C. As can be seen in Figure 3a,c, the surface of the as-received α + β alloy after immersion shows that the β phase is preferentially dissolved, while in Figure 3b, as-received α + β includes no significant attack of the α phase and not even in grain boundaries which are more susceptible to corrosion due to having less density and more disorder on the atomic scale. Also, they concluded that once the attack initiated in the two-phase alloy (as-received α + β), the β phase seems to be more susceptible to dissolution than the α phase. In addition, the significant conclusion was that the localized galvanic series of α and β phases causes preferential corrosion of the β phase due to the presence of vanadium (V) as a beta-stabilizer element in Ti-6Al-4V alloy, and this is the main reason for higher corrosion rate of Ti-6Al-4V in comparison with Ti-13Mo-7Zr-3Fe (TMZF) (Figure 3c) [40].

In a study by Lu et al. [45], the effect of α and β precipitations on the corrosion behaviour of β-type Ti-5Al-3Zr-4Mo-4Cr-4V alloy (Ti-1300) was investigated. Similar to the conclusion of [40], it was observed that both the equiaxed (the equiaxed microstructure of Ti-1300 alloy presents mainly relatively large equiaxed β-grains)/lamellar (the lamellar microstructure for Ti-1300 alloy is dominated by large quantities of precipitates as α phase and retained β phase) microstructures undergo a preferential corrosion degradation of α phase and the α + β interphase boundaries which create localized micro galvanic series between α and β phases (Figure 4a,b).

It is important to note that many previous investigations on the corrosion behaviour of β-type titanium alloys (for biomedical applications) focused on the alloy composition only in one solution. As discussed previously, pH is an important factor which varies in various solutions. Hence, every solution may impose a different electrochemical result on an individual alloy and studying different simulated body fluid solutions would better exhibit the role of solution components on the biomedical corrosion procedure in these alloys.

Cheng-hao et al. [46] studied the corrosion characteristics of Ti-6Al-4V alloy in different simulated body fluids including Ringer's solution (NaCl 8.5, KCl 0.2, $CaCl_2$ 0.2, $NaHCO_3$ 1.1 g/L), phosphate-buffered saline (PBS) solution (NaCl 8.0, KCl 0.2, Na_2HPO_4 1.15, KH_2PO_4 0.2 g/L) and Hank's solution (NaCl 0.8, KCl 0.4, $CaCl_2$ 0.14, $NaHCO_3$ 0.35, $C_6H_{12}O_6$ 1.0, $MgCl_2$. $6H_2O$ 0.1,

$MgSO_4 \cdot 7H_2O$ 0.06, Na_2HPO_4 0.06, KH_2PO_4 0.06 g/L) with different pH values. It was indicated that the order of corrosion rate (from highest to lowest) is in Ringer's solution, PBS solution and Hank's solution, respectively. It is reported that Na_2HPO_4 and KH_2PO_4 react with the metal surface and act as inhibitor by formation of a passive layer for the Hank's solution. Also, a further reduction of pH decreases the corrosion resistance of the alloy in all the solutions (Table 1).

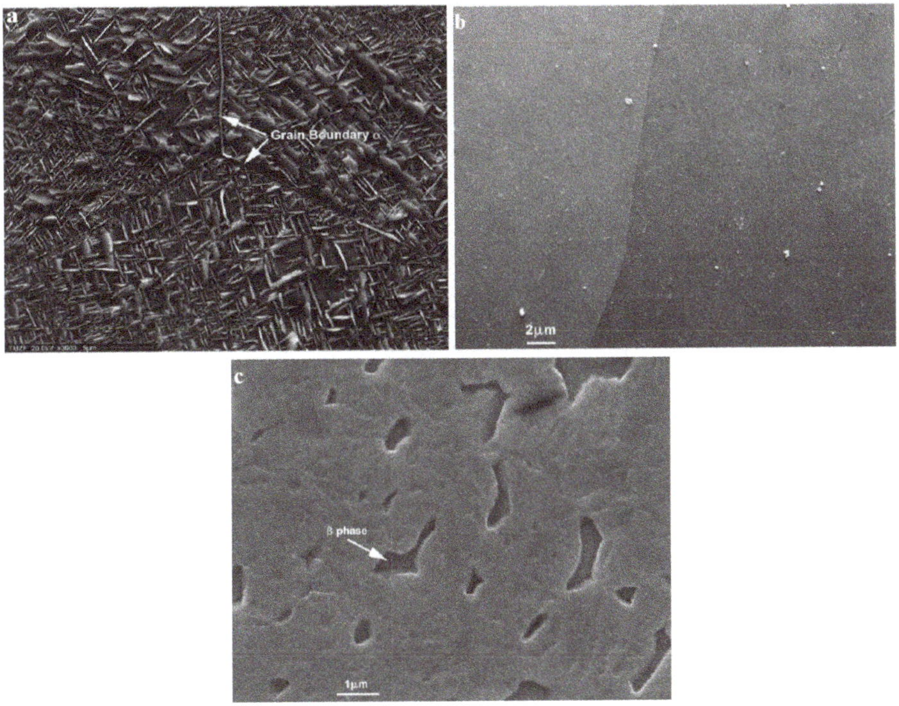

Figure 3. Secondary electron SEM micrograph of Ti-13Mo-7Zr-3Fe: (**a**) as-received $\alpha + \beta$, (**b**) metastable β, following 50 h immersion in 5 M HCl at 37 °C, and (**c**) secondary electron SEM micrograph of the surface corrosion aspects of Ti-6Al-4V ELI (extra-low interstitial) after 50 h immersion in 5 M HCl at 37 °C [40].

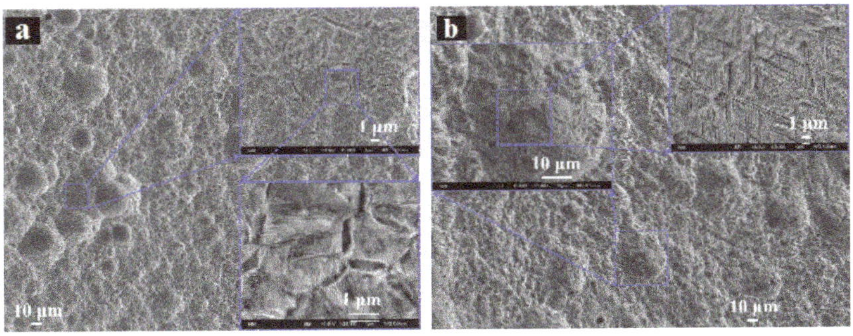

Figure 4. SEM morphologies of Ti-1300 (Ti-5Al-3Zr-4Mo-4Cr-4V) surface in 5 M HCl solution: (**a**) lamellar, and (**b**) equiaxed microstructure [45].

Table 1. Corrosion current densities in different simulated body fluids for Ti-6Al-4V [46].

Solution	PBS (pH 7.2)	Hank's (pH 7.2)	Ringer's		
			pH 7.2	pH 5.5	pH 3.5
Corrosion Current Density (nA/Cm^2)	31.31	24.8	38.13	58.5	76.45

Also, different simulated body fluids were used by Hasena et al. [47] in order to evaluate the electrochemical behaviour of NiTi alloy. Figure 5 exhibits the OCP (open circuit potential) of all the tested solutions. Higher Cl^- concentrations at 37 °C caused the most active potentials in HBSS (Hank's balanced salt solution), Ringer's solutions and the noblest (−291 mV) for HBSS after 3600 s exposure time. Also, it is claimed that in Hank's solution with pH ranging from 1 to 9, the corrosion rate is highly affected by a further decrease in pH [48]. Many authors [49–51], explained that the Ringer's solution seems to be more aggressive due to more concentrations of Cl^- ions. It is recommended to refer to Tas's work [52] for detailed reports on the effect of different ions on the corrosion rate of the tested alloys for each simulated body fluid solution.

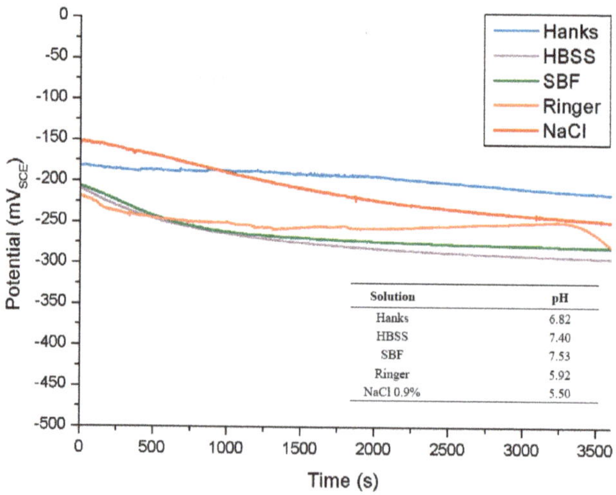

Figure 5. Open circuit potentials (OCP) in different simulated body fluids at 37 °C and in 0.9% NaCl at 25 °C [47].

3. Fabrication Process

The corrosion and electrochemical behaviour of two distinct alloys with the same chemical composition but different manufacturing processes may be different. Various heat treatments and mechanical processes of Ti alloys including ageing and cold working are expected to influence the microstructure and alloy distribution with a direct consequence on their corrosion behaviour. Also, surface treatments such as anodizing and coatings can inventively alter the electrical current density passing through the bulk material [53–55]. Yun Bai et al. [56] showed that wrought Ti-6Al-4V has greater corrosion rate than electron beam melting (EBM)-produced Ti-6Al-4V. EBM-produced Ti-6Al-4V mainly consists of α and β phases, with a higher volume of β phase and much refined lamellar α/β phases and it was found that localized micro galvanic cells produced by α and β phases degrade the corrosion resistance of EBM-produced Ti-6Al-4V. In a study by Buciumeanu et al. [57], three different fabrication processes including casting, powder metallurgy (hot pressing, HP) and laser engineered net shaping (LENS) were used for Ti-6Al-4V specimens. The OCP measurements of HP, LENS and cast Ti-6Al-4V alloy were around −0.25 V, −0.03 V and −0.36 V, respectively. Hence, the LENS samples were found to have a better corrosion protection than the other two processes. This characteristic was attributed to the very high and localized cooling rates during the LENS process resulting in higher hardness and slight chemistry variation in the material.

Potentiodynamic polarization tests were performed for two deformed Ti-15Zr-12Nb alloys including: warm rolling (WR) at 660 °C for 1 h up to the 95% thickness reduction, and hot rolling (HR) from 880 °C to 660 °C up to the 70% thickness reduction using a multi-pass rolling process [58]. It is indicated that the HR sample with a coarse lamellar structure has higher corrosion current density (0.9976 mA/cm^2) compared to the WR sample (0.8179 mA/cm^2) with an ultra-fine microstructure and formation of more stable and passive surface layer. Also, in a study by Li et al. [59], corrosion characterization of two Ti-24Nb-2Zr (at. %) samples were investigated. Both samples were 75% deformed and then annealed at 800 °C for 1 h, one of them was also aged at 300 °C for 1 h. The electrochemical tests were performed to make a comparison against CP Ti alloy. The both deformed samples had higher passive current densities than CP Ti due to the precipitation of ω phase during the deformation and ageing processes which significantly destabilize the passive oxide film [56]. According to the results of [58,59], not any deformation procedure can improve the corrosion resistance of β type Ti alloys. There are a range of parameters such as deformation percent, annealing temperature and the texture of refined grains (e.g., uniform orientation) can affect the electrochemical behaviour of the alloy.

Five treated Ti-29Nb-13Ta-4.6Zr specimens were tested [60] namely: 87.5% cold-rolled titanium-niobium-tantalum-zirconium ($TNTZ_{CR}$), solutionized at 790 °C for 1 h ($TNTZ_{ST}$), $TNTZ_{ST}$ aged at 400 °C for 72 h, multi-step-thermo-mechanical treatment ($TNTZ_{multi}$), where the cold rolling (CR) and solution treatment (ST) were repeated 4 times, $TNTZ_{multi}$ aged at 400 °C for 72 h. and hot-swaged Ti-30Nb-10Ta-5Zr, which has 0.2% O and mirror surface produced by buff polishing. Figure 6 shows a higher level of corrosion resistance for the aged ($TNTZ_{multi}$) and deformed ($TNTZ_{CR}$) samples when compared to the $TNTZ_{ST}$ and $TNTZ_{ST}$ aged at 400 °C. It is also shown that the critical current densities (I_c) and passive current densities (I_p) are at lowest for the $TNTZ_{multi}$ aged and $TNTZ_{multi}$ samples.

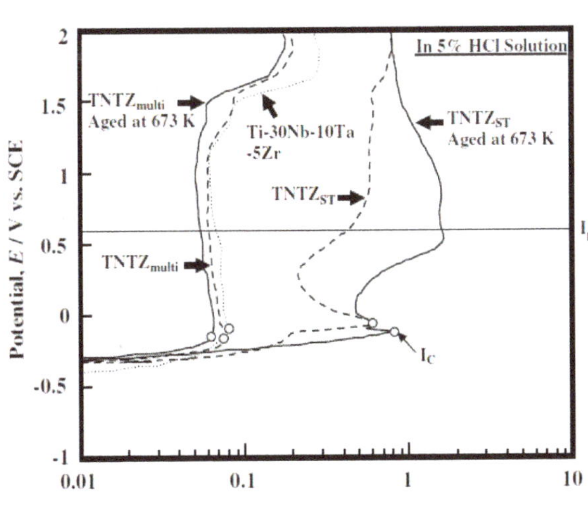

Figure 6. Anodic polarization curves of $TNTZ_{ST}$ (Ti-29Nb-13Ta-4.6Zr), $TNTZ_{ST}$ aged at 673 K for 259.2 ks $TNTZ_{multi}$, $TNTZ_{multi}$ aged at 673 K for 259.2 ks and hot swaged Ti-30Nb-10Ta-5Zr in 5% HCl solution at 310 K [60].

Two Ti-1.45Al-6.62Mo-4.53Fe-0.14O samples were solution treated at different temperatures of 750 °C (equiaxed β-grain structure with almost needle-like morphology of 10% primary α-phase) and

650 °C (equiaxed β-grain structure with almost needle-like morphology of 15% primary α-phase) for 30 min; and subsequently, both the samples were aged treated at 500 °C for 4 h. Through a comprehensive corrosion study, it was reported that the higher corrosion rate of 650 °C sample was due to the greater concentration of the α-phase which makes a local galvanic cell between the α and β phases [61]. It is important to note that thermal and mechanical treatments may alter the microstructure and phases in Ti alloys including β type Ti alloys. Therefore, the effect of these thermal and mechanical treatments on the corrosion behaviour of beta titanium alloys can be as important as the chemical composition. The corrosion current density reported in the literature [58] is (2×10^{-2} mA·cm^{-2}) for the 650 °C sample which is about 34 times greater than that of 750 °C (5.9×10^{-4} mA·cm^{-2}). It is thus indicated that a 100 °C increase in the solutionizing temperature of two equal alloys can greatly change the corrosion behaviour so that this might not happen even if the samples were made of two distinct precursors and initial composition. In another investigation [62], Ti-40Ta-22Hf-11.7Zr (TTHZ) beta Ti alloys were prepared in six samples, namely: as cast, ST (solution treated) and the other samples were then aged (STA) at 300 °C for 15 min, 1.5 h, 12 h and 24 h in argon, followed by air cooling (STA-15 min, STA-1.5 h, STA-12 h and STA-24 h), respectively. It was found that the ST sample has the best corrosion resistance consisting of a single beta phase, and with a further increase in ageing time, the α phase grows and the corrosion resistance decreases. Also, the STA-15 min sample had a small amount of α″ phase and β phase transformed into an α″ phase without diffusion, so the corrosion resistance of the alloy tended to be the same as that in the solid solution due to the close thermodynamic properties of α″ to β phase which is the middle state of transition of α to β. Nishimura et al. [63] explained that the ageing heat treatment created α phase precipitations in the β phase of solutionized Ti-10Mn alloys which act as anodic particles [63]. Finally, the dissolution priority for α and β phases seems to be different in their various combinations. The single α phase is less corrosion resistant than β phase. However, according to [37], once the attack initiated in the two-phase (as-received α + β) alloy, the β phase appeared to be less resistant to continued dissolution than the α phase.

4. Chemical Composition

Admittedly, in order to achieve a superior biocompatible medical implant material, a comprehensive corrosion study is needed on the electrochemical reactions of the material with the synthetic physiological solutions [64,65]. Like other alloys, the electrochemical behaviour of beta titanium alloys depends, to a great extent, on many parameters such as electrolyte, microstructure and importantly chemical composition of the alloy (alloying elements and their amounts) [63,66,67]. Chemical composition is a major parameter which can influence the corrosion resistance of titanium alloys including new beta type titanium alloys. For instance, small additions of indium (In) to a beta-type Ti-Nb alloy (tested in Ringer's solution) do not yield any negative effect on the corrosion behaviour and it was reported that a further addition of indium will enhance the corrosion resistance of the alloy [68]. Further addition of ruthenium (Ru) to Ti-20Nb-xRu (x = 0, 0.5, 1.0, 1.5 at. %) improves the corrosion resistance of this beta titanium alloy. The potentiodynamic polarization curves of Ru added samples in comparison to Ti-6Al-4V and CP Ti are presented in Figure 7. Approximately a 0.23 V difference in the corrosion potential of the Ru-free sample to nobler amounts and 1.6 V shift in comparison to the CP Ti and Ti-6Al-4V show a relatively better corrosion resistance [69].

An electrochemical study was performed on the corrosion behaviour of two beta-type titanium alloys namely Ti-15Mo (TiMo) and Ti-29Nb-13Ta-4.6Zr (TNTZ) [70]. After one hour of immersion in Hank's solution, the measured OCP for TiMo and TNTZ were −320 and −260 mV, respectively. This nobler potential of TNTZ beta titanium alloys related to the alloying elements increases the corrosion resistance. Using electrochemical impedance spectroscopy (EIS), the passivation behaviour of Ti-15Mo (TiMo) and Ti-29Nb-13Ta-4.6Zr (TNTZ) alloys was characterised. TNTZ showed two time constants at higher potentials, indicating the presence of two passive layers, namely an inner layer for corrosion resistance and an outer porous layer. Therefore, TNTZ seems to act better in orthopaedic applications [68]. Many authors have delved deeply in the electrochemical properties of low modulus

Ti alloys including beta phase [71–77]. The Ti passive layer acts as a protective layer in various aqueous environments and its reaction in H_2O is as follows:

$$Ti + H_2O \leftrightarrow Ti(OH)_4 + 2H_2$$

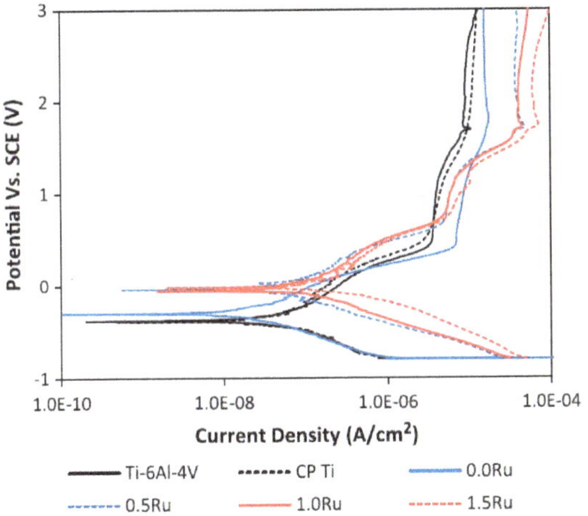

Figure 7. Representative potentiodynamic polarization curves of titanium alloys in Hank's balanced salt solution (HBSS) [69].

Many alloying elements in titanium alloys lead to a better corrosion and electrochemical behaviour, and this greatly depends on the pH value of the environment. The oxide layer properties formed on the surface of Ti alloys highly depend on the alloying elements present in the passive layer. Factors such as diffusion rate of the alloying element ions in the passive oxide film and mutual solubility can specify the strength and stability of the surface film on Ti alloys [78–82]. Metikoš-Huković et al. [82] showed that the alloying of titanium with elements of higher valence electrons such as Nb (Ti-6Al-4V vs. Ti-6Al-6Nb) can cause the stabilization of the passive film and elimination of anion vacancies. The presence of Nb^{5+} cations increases the number of oxygen ions, which cancel out the anion vacancies and make the film less defective [83]. The disappearance of the anion vacancies lowers the diffusion rates across the passive layer.

Also, the formation of intermetallic precipitates may provide susceptible sites for localized corrosion and stabilization of an α/β phase structure which may result in a lower homogeneity in the passive layer; and consequently, lowering the breakdown potential of the passive layer [84,85].

Titanium-zirconium-molybdenum (TZM), as a functional alloy is not only used in biomedical applications but also in many other industrial applications. Hence, the electrolyte used for corrosion investigations for this alloy is mostly ordinary NaCl and NaOH solutions [86–88]. Zirconium has excellent solid solubility in titanium due to its near electrochemical and mechanical aspects to titanium. Therefore, alloying these two elements can lead to an improvement in the corrosion and mechanical behaviour of the alloy [89,90]. Any increase of molybdenum may lead to a decrease in β-transus temperature, and Mo has a higher β stabilizing impact on the TZM than Zr [91].

A common version of TZM alloys is lanthanum-doped titanium-zirconium-molybdenum (La-TZM) alloy with considerable properties such as high temperature oxidation, and better mechanical and physical functions. This wide range of features of La-TZM alloy have encouraged many scientists to investigate its corrosion behaviour too [92–98]. Deng et al. [98] showed that alloying lanthanum in TZM degrades the corrosion resistance. Since the corrosion potential difference of La and other alloying elements in La-TZM alloy is high, the micro galvanic cell series created by this potential difference in the alloy matrix can easily cause the initiation of pits on the surface leading to intergranular corrosion. Lanthanum plays the role of the second-phase particle which includes a nobler potential than the Mo matrix and spoils the integration of the oxide passive layer on the surface. Also, in Figure 8, it is shown that no significant passivation behaviour in the anodic branch of La-TZM alloy occurs while the passivation behaviour of the TZM is obviously illustrated by the reduction in the current density around −4 A/cm^2. In another research [99], it is indicated that the corrosion resistance of TZM is better than La-TZM in alkaline and neutral environments including Cl^- and OH^-; while, on the contrary, the La-TZM shows a higher corrosion resistance than TZM in acidic media. Also, Hu et al. [99] indicated that the Cl^- concentration is of great importance to evaluate the corrosion behaviour of TZM alloy. They showed that TZM alloys have a good corrosion resistance in 0.5 mol/L or 1.5 mol/L Cl^- concentrations.

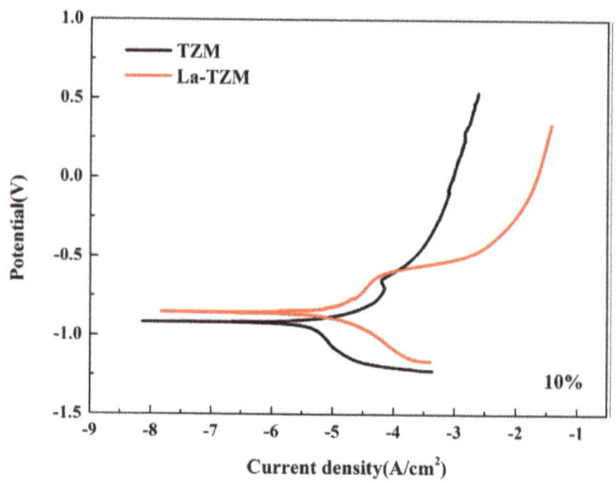

Figure 8. Potentiodynamic polarization curves of TZM and La-TZM alloys [98].

Quaternary TNTZ (titanium–niobium–tantalum–zirconium) alloys were developed in the 1990's [100]. These alloying elements improved the mechanical and corrosion behaviour of low modulus beta titanium alloys for biomedical applications. Nb is known as a β-stabilizer having a significant role in the spontaneous passivation of the alloy [101]. Also, from a microstructural point of view, a significant reduction of β-transus temperature seems to be present for the higher β-stabilizer content in some implants which include niobium in their composition [102,103]. Zr improves the passivity and lowers the active electrochemical reaction of the α phase of titanium [66] and Ta is considered as an element with an excellent capability in formation of passive oxide layer (Ta_2O_5) [104]. The most notable feature of TNTZ alloys, which improves its passive layer corrosion potential breakdown point in comparison with the other ternary and low modulus beta titanium alloys, is the strong bonding and adherence of their passive layers, near the corrosion potential of the alloying elements [105,106]. In a study reported by [107], it is shown that in Ti-20Nb-10Zr-5Ta alloy, Zr had the greatest influence on reducing the open circuit potential to more negative values and achieving a better protective passive layer behaviour. Milošev et al. [19] showed that addition of oxygen to TNT alloys may increase the chemical solubility of Ti due to its strong β-stabilizing effect and they tested the corrosion current density (0.045 μA/cm^2). Also, they reported that the incorporation of different alloying oxides in the surface layer is the main reason of improvement in the corrosion behaviour of Ti-20Nb-10Zr-5Ta alloys.

The other biomedical low modulus beta type titanium alloys of great interest are the ternary TNZ (Titanium-Niobium-Zirconium) alloys. Robin et al. [36] showed that the corrosion resistance of Ti-13Nb-13Zr alloy is a little higher than Ti-5Nb-13Zr and Ti-20Nb-13Zr alloys in the Ringer's solution excluding fluorides, and discussed that this is due to the lower Nb content of α-phase in

Ti-13Nb-13Zr. In a TNZ system, Nb and Zr encourage the suppression of O_2 evolution and cease the cathodic electron transfer reaction and can lower the active anodic dissolution reactions in the system [108]. Huang et al. [109] compared two alloys of Ti-13Nb-13Zr and Ti-15Mo, and concluded that Ti-13Nb-13Zr has a superior anticorrosion performance because of the Nb alloying element which annihilates the anion vacancies in the crystal lattice of the TiO_2 passive layer, which is due to the formation of covalent bonds among Nb, Zr and Ti (4d3, 4d2, 3d2) respectively [66]. Also, Macdonald [110] presented the theory of the point defect model in surface layer whose presence can cause heterogeneity of passive layer and degradation of corrosion resistance [110]. Hence, Nb^{5+} particles can situate in the crystal lattice of TiO_2 and fill the anion vacancies which results in lower defects and higher corrosion resistance [82].

An important ternary and quaternary group of beta titanium alloys including Sn alloying element has emerged recently. A study on Ti-32Nb-2Sn and Ti-32Nb-4Sn concluded that the higher Sn content improves the corrosion behaviour due to the higher passive layer quality [110]. Also, it must be considered that the Sn content up to 4 wt% to 6 wt% may cause cytotoxic effects in the human body and, therefore, further investigations should be undertaken [111]. Bai et al. [2] reported that Sn included beta titanium alloys show a wide range of passivation on anodic branch of Tafel diagram and it is a superior feature of these alloys.

5. General Significations of Corrosion

The complicated electrolytic behaviour of body fluid including a combination of attacking corrosion components creates an aggressive media for biomedical implants. Hence, controlling and improving the corrosion resistance of medical implants seems to be a significant concern in the manufacturing process of implants [112]. In spite of electrochemical parameters, mechanical deformation affects the corrosion rate as seen in [113] where a further amount of cold rolling lowered the corrosion rate. Thermodynamics and kinetics are the main two features of the electrochemical corrosion phenomenon. Thermodynamics specifies the oxidation-reduction reactions of the implant material adjacent to the body fluid media and the kinetics evaluates the probability of occurrence of reactions [114]. Common types of corrosion that can occur in beta type titanium alloys (when used in implants) are fretting, galvanic, pitting/crevice and uniform corrosion [115]. In an implant such as hip joint implants of which all the surface is in contact with the body tissues and fluids, the uniform corrosion is of great concern and pH variations can play a significant role in the corrosion degradation. Also, surface treatments and the thermodynamic stability and high adherence of the passive layer may control this type of corrosion [116]. In addition, fretting and wear corrosion must be considered in modular implants where metallic parts have direct contact (e.g., taper junction of hip implants). With regard to the advancement and development of low modulus beta titanium alloys, it is expected that successful implant materials withstand long-term uniform corrosion with corrosion rate of less than 1 μm/year [117]. For dental implants, pitting and galvanic corrosion should be more investigated since food debris is always deposited and can make a localized galvanic series following pitting corrosion [118]. Moreover, it is important to perform microstructural evaluations along with ordinary corrosion testing procedures when investigating pitting and galvanic corrosion in dental implant materials. Surface images of dental implants and any other implants susceptible to pitting corrosion can provide useful data about the morphology, distribution, depth and number of defects. Hence, further experiments would be more reliable and authentic.

6. Conclusions

In this review, low modulus β-type titanium alloys were investigated from a corrosion behaviour point of view. The important parameters that can affect the electrochemical and corrosion behaviour of these alloys were discussed. Investigations on the development of a suitable microstructure with optimal mechanical properties have been performed to design and fabricate low modulus β-type Ti-based alloys for medical implant applications. However, less attention has been paid to the electrochemical and corrosion behaviour of these new generation titanium alloys. Toxicity of biomedical implants depends, to a great extent, on the ion release rate of the alloy into the body which passes through the surface passive layer. Hence, the microstructure and composition of the passive layer and its adhesion and stability should be carefully evaluated in these alloys. Also, surface treatments and anodizing can greatly improve the quality of passive layers but the chemical composition of the surface coatings should be non-toxic. Surface microstructural observations of beta type Ti alloys after electrochemical tests in various simulated body fluid solutions play an important role in verifying corrosion test results. These observations can also reveal more data as to other corrosion aspects of the surface such as thickness of the passive layer, corroded phases and particles, pitting attacked areas and corrosion mechanism of the bulk. It is important to note that the physical and mechanical changes in beta-type Ti alloys may alter the microstructure and phases. Therefore, the effect of these physical (heat) and mechanical treatments on the corrosion behaviour of beta titanium alloys may be as important as the chemical composition. Also, the amount of α and β phases in the microstructure and their interaction and dissolution can affect the corrosion resistance. It is important to mention that various physical and mechanical treatments of beta titanium alloys directly affect the interaction and distribution of α and β phases which these phases not only alter the corrosion behaviour in the bulk material but also play an important role on the surface passive layer of implants.

Dissolution priority for the α and *β* phases seems to be different in their various combinations. The single α phase is less corrosion resistant than β phase. However, once the attack has initiated in the two-phase (α + β), the β phase appears to be less resistant to continued dissolution when compared to the α phase. It is suggested that more investigations should be performed on the corrosion resistance of α and β phases (separately and as mixed). Suitable amounts of α and β phases should be created in the microstructure of beta-titanium alloys including various alloying elements in order to have a high corrosion resistance and a stable surface passive layer. Afterwards, other physical, mechanical and chemical treatments such as surface treatment, heat treatments (e.g., ageing, solutionizing, and hardening) and various alloy fabrication processes should be taken into account.

Author Contributions: Conceptualization, R.H.O.; literature review, P.A.; writing—original draft preparation, P.A.; writing—review and editing, R.O. and R.G.; supervision, R.O. and R.G.

Funding: This research received no external funding.

Conflicts of Interest: The authors declare no conflict of interest.

References

1. Kunčická, L.; Kocich, R.; Lowe, T.C. Advances in metals and alloys for joint replacement. *Prog. Mater. Sci.* **2017**, *88*, 232–280.
2. Bai, Y.; Li, S.J.; Prima, F.; Hao, Y.L.; Yang, R. Electrochemical corrosion behavior of Ti-24Nb-4Zr-8Sn alloy in a simulated physiological environment. *Appl. Surf. Sci.* **2012**, *258*, 4035–4040. [CrossRef]
3. Zhou, Y.L.; Niinomi, M.; Akahori, T.; Fukui, H.; Toda, H. Corrosion resistance and biocompatibility of Ti-Ta alloys for biomedical applications. *Mater. Sci. Eng. A* **2005**, *398*, 28–36. [CrossRef]
4. Yamamuro, T. Patterns of osteogenesis in relation to various biomaterials. *J. Jpn. Soc. Biomater.* **1989**, *7*, 19–23.
5. Niinomi, M.; Nakai, M.; Hieda, J. Development of new metallic alloys for biomedical applications. *Acta Biomater.* **2012**, *8*, 3888–3903. [CrossRef] [PubMed]
6. Hua, N.; Chen, W.; Zhang, L.; Li, G.; Liao, Z.; Lin, Y. Mechanical properties and bio-tribological behaviors of novel beta-Zr-type Zr-Al-Fe-Nb alloys for biomedical applications. *Mater. Sci. Eng. C* **2017**, *76*, 1154–1165. [CrossRef] [PubMed]
7. Liu, H.; Niinomi, M.; Nakai, M.; Obara, S.; Fujii, H. Improved fatigue properties with maintaining low young's modulus achieved in biomedical beta-type titanium alloy by oxygen addition. *Mater. Sci. Eng. A* **2017**, *704*, 10–17. [CrossRef]
8. Manam, N.S.; Harun, W.S.W.; Shri, D.N.A.; Ghani, S.A.C.; Kurniawan, T.; Ismail, M.H.; Ibrahim, M.H.I. Study of corrosion in biocompatible metals for implants: A review. *J. Alloys Compd.* **2017**, *701*, 698–715. [CrossRef]

9. Akahori, T.; Niinomi, M.; Fukui, H.; Ogawa, M.; Toda, H. Improvement in fatigue characteristics of newly developed beta type titanium alloy for biomedical applications by thermo-mechanical treatments. *Mater. Sci. Eng. C* **2005**, *25*, 248–254. [CrossRef]
10. Du, Z.; Xiao, S.; Liu, J.; Lv, S.; Xu, L.; Kong, F.; Chen, Y. Hot deformation behavior of Ti-3.5Al-5Mo-6V-3Cr-2Sn-0.5Fe Alloy in $\alpha + \beta$ field. *Metals* **2015**, *5*, 216–227. [CrossRef]
11. Biesiekierski, A.; Wang, J.; Gepreel, M.A.H.; Wen, C. A new look at biomedical Ti-based shape memory alloys. *Acta Biomater.* **2012**, *8*, 1661–1669. [CrossRef] [PubMed]
12. Oskouei, R.; Fallahnezhad, K.; Kuppusami, S. An investigation on the wear resistance and fatigue behaviour of Ti-6Al-4V notched members coated with hydroxyapatite coatings. *Materials* **2016**, *9*, 111. [CrossRef] [PubMed]
13. Raducanu, D.; Vasilescu, E.; Cojocaru, V.D.; Cinca, I.; Drob, P.; Vasilescu, C.; Drob, S.I. Mechanical and corrosion resistance of a new nanostructured Ti-Zr-Ta-Nb alloy. *J. Mech. Behav. Biomed. Mater.* **2011**, *4*, 1421–1430. [CrossRef] [PubMed]
14. Niinomi, M. Mechanical biocompatibilities of titanium alloys for biomedical applications. *J. Mech. Behav. Biomed. Mater.* **2008**, *1*, 30–42. [CrossRef] [PubMed]
15. Yilmazer, H.; Niinomi, M.; Nakai, M.; Hieda, J.; Todaka, Y.; Akahori, T.; Miyazaki, T. Heterogeneous structure and mechanical hardness of biomedical β-type Ti-29Nb-13Ta-4.6Zr subjected to high-pressure torsion. *J. Mech. Behav. Biomed. Mater.* **2012**, *10*, 235–245. [CrossRef] [PubMed]
16. Niinomi, M.; Hattori, T.; Morikawa, K.; Kasuga, T.; Suzuki, A.; Fukui, H.; Niwa, S. Development of low rigidity β-type titanium alloy for biomedical applications. *Mater. Trans.* **2002**, *43*, 2970–2977.
17. Takematsu, E.; Cho, K.; Hieda, J.; Nakai, M.; Katsumata, K.; Okada, K.; Niinomi, M.; Matsushita, N. Adhesive strength of bioactive oxide layers fabricated on TNTZ alloy by three different alkali-solution treatments. *J. Mech. Behav. Biomed. Mater.* **2016**, *61*, 174–181. [CrossRef]
18. Bahl, S.; Das, S.; Suwas, S.; Chatterjee, K. Engineering the next-generation tin containing β titanium alloys with high strength and low modulus for orthopedic applications. *J. Mech. Behav. Biomed. Mater.* **2018**, *78*, 124–133. [CrossRef]
19. Milošev, I.; Žerjav, G.; Moreno, J.M.C.; Popa, M. Electrochemical properties, chemical composition and thickness of passive film formed on novel Ti-20Nb-10Zr-5Ta alloy. *Electrochim. Acta* **2013**, *99*, 176–189. [CrossRef]
20. Brett, C.; Muresan, I. The influence of artificial body fluids on metallic corrosion. In *Key Engineering Materials*; Trans Tech Publications: Zurich, Switzerland, 2002; Volume 230, pp. 459–462.
21. Fallahnezhad, K.; Oskouei, R.H.; Badnava, H.; Taylor, M. An adaptive finite element simulation of fretting wear damage at the head-neck taper junction of total hip replacement: The role of taper angle mismatch. *J. Mech. Behav. Biomed. Mater.* **2017**, *75*, 58–67. [CrossRef]
22. Oskouei, R.H.; Barati, M.R.; Farhoudi, H.; Taylor, M.; Solomon, L.B. A new finding on the in-vivo crevice corrosion damage in a CoCrMo hip implant. *Mater. Sci. Eng. C* **2017**, *79*, 390–398. [CrossRef] [PubMed]
23. Huang, G.Y.; Jiang, H.B.; Cha, J.Y.; Kim, K.M.; Hwang, C.J. The effect of fluoride-containing oral rinses on the corrosion resistance of titanium alloy (Ti-6Al-4V). *Korean J. Orthod.* **2017**, *47*, 306–312. [CrossRef] [PubMed]
24. Fallahnezhad, K.; Oskouei, R.H.; Taylor, M. Development of a fretting corrosion model for metallic interfaces using adaptive finite element analysis. *Finite Elem. Anal. Des.* **2018**, *148*, 38–47. [CrossRef]
25. Fallahnezhad, K.; Farhoudi, H.; Oskouei, R.H.; Taylor, M. A finite element study on the mechanical response of the head-neck interface of hip implants under realistic forces and moments of daily activities: Part 2. *J. Mech. Behav. Biomed. Mater.* **2018**, *77*, 164–170. [CrossRef] [PubMed]
26. Fallahnezhad, K.; Farhoudi, H.; Oskouei, R.H.; Taylor, M. A finite element study on the mechanical response of the head-neck interface of hip implants under realistic forces and moments of daily activities: Part 1. *J. Mech. Behav. Biomed. Mater.* **2017**, *75*, 470–476. [CrossRef] [PubMed]
27. Hussenbocus, S.; Kosuge, D.; Solomon, L.B.; Howie, D.W.; Oskouei, R.H. Head-neck taper corrosion in hip arthroplasty. *Biomed Res. Int.* **2015**, *2015*, 758123. [CrossRef] [PubMed]
28. Milimonfared, R.; Oskouei, R.; Taylor, M.; Solomon, L. The distribution and severity of corrosion damage at eight distinct zones of metallic femoral stem implants. *Metals* **2018**, *8*, 13–24. [CrossRef]
29. Milimonfared, R.; Oskouei, R.H.; Taylor, M.; Solomon, L.B. An intelligent system for image-based rating of corrosion severity at stem taper of retrieved hip replacement implants. *Med. Eng. Phys.* **2018**, *61*, 13–24. [CrossRef] [PubMed]

30. Streitzel, R.; Hösch, A.; Kalbfleisch, H.; Buch, D. In vitro corrosion of titanium. *Biomaterials* **1998**, *19*, 1495–1499. [CrossRef]
31. Schmidt, A.M.; Azambuja, D.S. Electrochemical behavior of Ti and Ti6Al4V in aqueous solutions of citric acid containing halides. *Mater. Res.* **2006**, *9*, 387–392. [CrossRef]
32. Schutz, R.W.; Thomas, D.E. Corrosion of titanium and titanium alloys. *Metals Handb. Corros.* **1987**, *13*, 669–706.
33. Reclaru, L.; Meyer, J.M. Effects of fluorides on titanium and other dental alloys in dentistry. *Biomaterials* **1998**, *19*, 85–92. [CrossRef]
34. Li, J.; Bai, Y.; Fan, Z.; Li, S.; Hao, Y.; Yang, R.; Gao, Y. Effect of fluoride on the corrosion behavior of nanostructured Ti-24Nb-4Zr-8Sn alloy in acidulated artificial saliva. *J. Mater. Sci. Technol.* **2018**, *34*, 1660–1670. [CrossRef]
35. Huang, H.H. Effect of fluoride and albumin concentration on the corrosion behavior of Ti-6Al-4V alloy. *Biomaterials* **2003**, *24*, 275–282. [CrossRef]
36. Robin, A.; Carvalho, O.A.S. Influence of pH and fluoride species on the corrosion behavior of ti-xnb-13zr alloys in ringer's solution. *Adv. Mater. Sci. Eng.* **2013**, *2013*, 434975. [CrossRef]
37. Reyes, K.M.; Kuromoto, N.K.; Claro, A.P.R.A.; Marino, C.E.B. Electrochemical stability of binary TiNb for biomedical applications. *Mater. Res. Express* **2017**, *4*, 075402. [CrossRef]
38. Prashanth, K.; Zhuravleva, K.; Okulov, I.; Calin, M.; Eckert, J.; Gebert, A. Mechanical and corrosion behavior of new generation Ti-45Nb porous alloys implant devices. *Technologies* **2016**, *4*, 33. [CrossRef]
39. Fojt, J.; Joska, L.; Málek, J. Corrosion behaviour of porous Ti–39Nb alloy for biomedical applications. *Corros. Sci.* **2013**, *71*, 78–83. [CrossRef]
40. Atapour, M.; Pilchak, A.L.; Frankel, G.S.; Williams, J.C. Corrosion behavior of β titanium alloys for biomedical applications. *Mater. Sci. Eng. C* **2011**, *31*, 885–891. [CrossRef]
41. Samuel, S.; Nag, S.; Nasrazadani, S.; Ukirde, V.; Bouanani, M.E.; Mohandas, A.; Nguyen, K.; Banerjee, R. Corrosion resistance and in vitro response of laser-deposited Ti-Nb-Zr-Ta alloys for orthopedic implant applications. *J. Biomed. Mater. Res. Part A* **2010**, *94*, 1251–1256. [CrossRef] [PubMed]
42. Moreno, J.M.C.; Vasilescu, C.; Drob, S.I.; Popa, M.; Drob, P.; Vasilescu, E. Electrodeposition, characterization, and corrosion stability of nanostructured anodic oxides on new Ti-15Zr-5Nb alloy surface. *J. Nanomater.* **2013**, *2013*, 3.
43. Bai, Y.; Hao, Y.L.; Li, S.J.; Hao, Y.Q.; Yang, R.; Prima, F. Corrosion behavior of biomedical Ti-24Nb-4Zr-8Sn alloy in different simulated body solutions. *Mater. Sci. Eng. C* **2013**, *33*, 2159–2167. [CrossRef] [PubMed]
44. Jin, L.; Cui, W.F.; Song, X.; Liu, G.; Zhou, L. Effects of surface nanocrystallization on corrosion resistance of β-type titanium alloy. *Trans. Nonferr. Metals Soc. China* **2014**, *24*, 2529–2535. [CrossRef]
45. Lu, J.; Ge, P.; Li, Q.; Zhang, W.; Huo, W.; Hu, J.; Zhang, Y.; Zhao, Y. Effect of microstructure characteristic on mechanical properties and corrosion behavior of new high strength Ti-1300 beta titanium alloy. *J. Alloys Compd.* **2017**, *727*, 1126–1135. [CrossRef]
46. Liang, C.; Mou, Z. Effect of different simulated body fluids on anti-corrosion bimetallic materials. *Trans. Nonferr. Met. Soc. China* **2001**, *11*, 579–582.
47. Hansen, A.W.; Führ, L.T.; Antonini, L.M.; Villarinho, D.J.; Marino, C.E.B.; Malfatti, C.D.F. The electrochemical behavior of the NiTi alloy in different simulated body fluids. *Mater. Res.* **2015**, *18*, 184–190. [CrossRef]
48. Trépanier, C.; Pelton, A.R. Effect of temperature and pH on the corrosion resistance of nitinol. In medical device materials. In Proceedings of the Materials and Processes for Medical Devices Conference, Baden-Baden, Germany, 3–7 October 2004.
49. Shahrabi, T.; Sanjabi, S.; Saebnoori, E.; Barber, Z.H. Extremely high pitting resistance of NiTi shape memory alloy thin film in simulated body fluids. *Mater. Lett.* **2008**, *62*, 2791–2794. [CrossRef]
50. Liang, C.-H.; Mou, Z.-Q. Effects of different simulated fluids on anticorrosion biometallic materials. *Trans. Nonferr. Met. Soc. China* **2001**, *11*, 579–582.
51. Li, X.; Wang, J.; Han, E.H.; Ke, W. Influence of fluoride and chloride on corrosion behavior of NiTi orthodontic wires. *Acta Biomater.* **2007**, *3*, 807–815. [CrossRef]
52. Tas, A.C. The use of physiological solutions or media in calcium phosphate synthesis and processing. *Acta Biomater.* **2014**, *10*, 1771–1792. [CrossRef]
53. Afzali, P.; Yousefpour, M.; Borhani, E. Evaluation of the effect of ageing heat treatment on corrosion resistance of Al-Ag alloy using electrochemical methods. *J. Mater. Res.* **2016**, *31*, 2457–2464. [CrossRef]

54. Lopes, C.S.D.; Donato, M.T.; Ramgi, P. Comparative corrosion behaviour of titanium alloys (TI-15MO and TI-6AL-4V) for dental implants applications: A review. *Corros. Prot. Mater.* **2016**, *35*, 2.
55. Geetha, M.; Mudali, U.K.; Gogia, A.K.; Asokamani, R.; Raj, B. Influence of microstructure and alloying elements on corrosion behavior of Ti-13Nb-13Zr alloy. *Corros. Sci.* **2004**, *46*, 877–892. [CrossRef]
56. Bai, Y.; Gai, X.; Li, S.; Zhang, L.C.; Liu, Y.; Hao, Y.; Zhang, X.; Yang, R.; Gao, Y. Improved corrosion behaviour of electron beam melted Ti-6Al–4V alloy in phosphate buffered saline. *Corros. Sci.* **2017**, *123*, 289–296. [CrossRef]
57. Buciumeanu, M.; Bagheri, A.; Shamsaei, N.; Thompson, S.M.; Silva, F.S.; Henriques, B. Tribocorrosion behavior of additive manufactured Ti-6Al-4V biomedical alloy. *Tribol. Int.* **2018**, *119*, 381–388. [CrossRef]
58. Mohammed, M.T. Development of a new metastable beta titanium alloy for biomedical applications. *Karbala Int. J. Mod. Sci.* **2017**, *3*, 224–230. [CrossRef]
59. Li, Q.; Li, J.; Ma, G.; Liu, X.; Pan, D. Influence of ω phase precipitation on mechanical performance and corrosion resistance of Ti–Nb–Zr alloy. *Mater. Des.* **2016**, *111*, 421–428. [CrossRef]
60. Akahori, T.; Niinomi, M.; Fukui, H.; Suzuki, A. Fatigue, fretting fatigue and corrosion characteristics of biocompatible beta type titanium alloy conducted with various thereto-mechanical treatments. *Mater. Trans.* **2004**, *45*, 1540–1548. [CrossRef]
61. Ibrahim, K.M.; Moustafa, M.M.; Al-Grafi, M.W.; El-Bagoury, N.; Amin, M.A. Effect of solution heat treatment on microstructure and wear and corrosion behavior of a two phase β-metastable titanium alloy. *Int. J. Electrochem. Sci.* **2016**, *11*, 3206–3226. [CrossRef]
62. Lin, J.; Ozan, S.; Munir, K.; Wang, K.; Tong, X.; Li, Y.; Li, G.; Wen, C. Effects of solution treatment and aging on the microstructure, mechanical properties, and corrosion resistance of a β type Ti-Ta-Hf-Zr alloy. *RSC Adv.* **2017**, *7*, 12309–12317. [CrossRef]
63. Nishimura, T. Effect of microstructure on the electrochemical behavior of Ti-10 Mass% Mn alloys in high chloride solution. *J. Mater. Eng. Perform.* **2016**, *25*, 443–450. [CrossRef]
64. Mary, J.S.; Rajendran, S. Corrosion behaviour of metals in artificial body fluid an over view. *Zaštita Materijala* **2012**, *53*, 181–189.
65. Nagalakshmi, R.; Rajendran, S.; Sathiyabama, J.; Pandiarajan, M.; Christy, J.L. Corrosion behaviour of biomaterials in synthetic biological solutions—An overview. *Eur. Chem. Bull.* **2013**, *2*, 171–179.
66. Yu, S.Y.; Scully, J.R. Corrosion and passivity of Ti-13% Nb-13% Zr in comparison to other biomedical implant alloys. *Corrosion* **1997**, *53*, 965–976. [CrossRef]
67. Reyes, K.M.; Kuromoto, N.K.; Claro, A.P.R.A.; Marino, C.E.B. Potentiodynamic polarization study of type 316L and 316LVM stainless steels for surgical implants in simulated body fluids. *J. Chem. Pharm. Res.* **2012**, *4*, 203–208.
68. Gebert, A.; Oswald, S.; Helth, A.; Voss, A.; Gostin, P.F.; Rohnke, M.; Janek, J.; Calin, M.; Eckert, J. Effect of indium (In) on corrosion and passivity of a beta-type Ti-Nb alloy in Ringer's solution. *Appl. Surf. Sci.* **2015**, *335*, 213–222. [CrossRef]
69. Biesiekierski, A.; Ping, D.H.; Yamabe-Mitarai, Y.; Wen, C. Impact of ruthenium on microstructure and corrosion behavior of β-type Ti-Nb-Ru alloys for biomedical applications. *Mater. Des.* **2014**, *59*, 303–309. [CrossRef]
70. Jugowiec, D.; Łukaszczyk, A.; Cieniek, Ł.; Kowalski, K.; Rumian, Ł.; Pietryga, K.; Kot, M.; Pamuła, E.; Moskalewicz, T. Influence of the electrophoretic deposition route on the microstructure and properties of nano-hydroxyapatite/chitosan coatings on the Ti-13Nb-13Zr alloy. *Surf. Coat. Technol.* **2017**, *324*, 64–79. [CrossRef]
71. Karthega, M.; Raman, V.; Rajendran, N. Influence of potential on the electrochemical behaviour of β titanium alloys in Hank's solution. *Acta Biomater.* **2007**, *3*, 1019–1023. [CrossRef]
72. Bolat, G.; Mareci, D.; Chelariu, R.; Izquierdo, J.; González, S.; Souto, R.M. Investigation of the electrochemical behaviour of TiMo alloys in simulated physiological solutions. *Electrochim. Acta* **2013**, *113*, 470–480. [CrossRef]
73. Mareci, D.; Chelariu, R.; Sutiman, D.; Gordin, D.M.; Gloriant, T. Evaluating electrochemical behaviour of recrystallized titanium alloys in Ringer's solution. *Mater. Corros.* **2011**, *62*, 1117–1123. [CrossRef]
74. Mareci, D.; Chelariu, R.; Gordin, D.M.; Romas, M.; Sutiman, D.; Gloriant, T. Effect of Mo content on electrochemical behaviour of TiMo alloys for dental applications. *Mater. Corros.* **2010**, *61*, 829–837. [CrossRef]

75. Mareci, D.; Chelariu, R.; Bolat, G.; Cailean, A.; Grancea, V.; Sutiman, D. Electrochemical behaviour of Ti alloys containing Mo and Ta as β-stabilizer elements for dental application. *Trans. Nonferr. Met. Soc. China* **2013**, *23*, 3829–3836. [CrossRef]
76. Lee, Y.R.; Han, M.K.; Kim, M.K.; Moon, W.J.; Song, H.J.; Park, Y.J. Effect of gold addition on the microstructure, mechanical properties and corrosion behavior of Ti alloys. *Gold Bull.* **2014**, *47*, 153–160. [CrossRef]
77. Cremasco, A.; Messias, A.D.; Esposito, A.R.; Duek, E.A.D.R.; Caram, R. Effects of alloying elements on the cytotoxic response of titanium alloys. *Mater. Sci. Eng. C* **2011**, *31*, 833–839. [CrossRef]
78. Jablokov, V.R.; Nutt, M.J.; Richelsoph, M.E.; Freese, H.L. The application of Ti-15Mo beta titanium alloy in high strength structural orthopaedic applications. *J. ASTM Int.* **2005**, *2*, 1–8. [CrossRef]
79. Oliveira, N.T.C.; Guastaldi, A.C. Electrochemical stability and corrosion resistance of Ti-Mo alloys for biomedical applications. *Acta Biomater.* **2009**, *5*, 399–405. [CrossRef] [PubMed]
80. González, J.E.G.; Mirza-Rosca, J.C. Study of the corrosion behavior of titanium and some of its alloys for biomedical and dental implant applications. *J. Electroanal. Chem.* **1999**, *471*, 109–115. [CrossRef]
81. Guillemot, F. Recent advances in the design of titanium alloys for orthopedic applications. *Expert Rev. Med. Dev.* **2005**, *2*, 741–748. [CrossRef] [PubMed]
82. Metikoš-Huković, M.; Kwokal, A.; Piljac, J. The influence of niobium and vanadium on passivity of titanium-based implants in physiological solution. *Biomaterials* **2003**, *24*, 3765–3775. [CrossRef]
83. Kubaschewski, O.; Hopkins, B.E. *Oxidation of Metals*; Kubaschewski, V.O., Hopkins, B.E., Eds.; Butterworths Scientific Publications: London, UK, 1998.
84. Watanabe, T.; Shindo, T.; Naito, H. Effect of iron content on the breakdown potential for pitting of titanium in NaCl solutions. In *Sixth World Conference on Titanium IV*; Cedex: Paris, France, 1988; pp. 1735–1740.
85. Shoesmith, D.W.; Ikeda, B.M. The resistance of titanium to pitting, microbially induced corrosion and corrosion under unsaturated conditions. *Atomic Energy Canada* **1997**, *30*, 96–557.
86. Banerjee, D.; Williams, J.C. Perspectives on titanium science and technology. *Acta Mater.* **2013**, *61*, 844–879. [CrossRef]
87. Li, Y.; Yang, C.; Zhao, H.; Qu, S.; Li, X.; Li, Y. New developments of ti-based alloys for biomedical applications. *Materials* **2014**, *7*, 1709–1800. [CrossRef] [PubMed]
88. Ye, Y.; Liu, Z.; Liu, W.; Zhang, D.; Zhao, H.; Wang, L.; Li, X. Superhydrophobic oligoaniline-containing electroactive silica coating as pre-process coating for corrosion protection of carbon steel. *Chem. Eng. J.* **2018**, *348*, 940–951. [CrossRef]
89. Ho, W.F.; Cheng, C.H.; Pan, C.H.; Wu, S.C.; Hsu, H.C. Structure, mechanical properties and grindability of dental Ti-10Zr-X alloys. *Mater. Sci. Eng. C* **2009**, *29*, 36–43. [CrossRef]
90. Hsu, H.-C.; Wu, S.-C.; Sung, Y.-C.; Ho, W.-F. The structure and mechanical properties of as-cast Zr–Ti alloys. *J. Alloys Compd.* **2009**, *488*, 279–283. [CrossRef]
91. Correa, D.R.N.; Vicente, F.B.; Araújo, R.O.; Lourenço, M.L.; Kuroda, P.A.B.; Buzalaf, M.A.R.; Grandini, C.R. Effect of the substitutional elements on the microstructure of the Ti-15Mo-Zr and Ti-15Zr-Mo systems alloys. *J. Mater. Res. Technol.* **2015**, *4*, 180–185. [CrossRef]
92. Yang, F.; Wang, K.S.; Hu, P.; He, H.C.; Kang, X.Q.; Wang, H.; Liu, R.Z.; Volinsky, A.A. La doping effect on TZM alloy oxidation behavior. *J. Alloys Compd.* **2014**, *593*, 196–201. [CrossRef]
93. Hu, P.; Hu, B.L.; Wang, K.S.; Song, R.; Yang, F.; Yu, Z.T.; Tan, J.F.; Cao, W.C.; Liu, D.X.; An, G.; et al. Strengthening and elongation mechanism of Lanthanum-doped Titanium-Zirconium-Molybdenum alloy. *Mater. Sci. Eng. A* **2016**, *678*, 315–319. [CrossRef]
94. Hu, P.; Yang, F.; Deng, J.; Chang, T.; Hu, B.; Tan, J.; Wang, K.; Cao, W.; Feng, P.; Yu, H. High temperature mechanical properties of TZM alloys under different lanthanum doping treatments. *J. Alloys Compd.* **2017**, *711*, 64–70. [CrossRef]
95. Hu, P.; Zhou, Y.; Chang, T.; Yu, Z.; Wang, K.; Yang, F.; Hu, B.; Cao, W.; Yu, H. Investigation on compression behavior of TZM and La2O3doped TZM Alloys at high temperature. *Mater. Sci. Eng. A* **2017**, *687*, 276–280. [CrossRef]
96. Wang, K.S.; Tan, J.F.; Hu, P.; Yu, Z.T.; Yang, F.; Hu, B.L.; Song, R.; He, H.C.; Volinsky, A.A. La2O3 effects on TZM alloy recovery, recrystallization and mechanical properties. *Mater. Sci. Eng. A* **2015**, *636*, 415–420. [CrossRef]
97. Hu, B.L.; Wang, K.S.; Hu, P.; Zhou, Y.H.; Yang, F.; Li, Q.W.; Yu, H.L. Effect of lanthanum in arc erosion of Titanium-Zirconium-Molybdenum alloy. *J. Alloys Compd.* **2017**, *696*, 522–528. [CrossRef]

98. Deng, J.; Wang, K.S.; Hu, P.; Zhou, Y.H.; Chang, T.; Hu, B.L.; Feng, P.F.; Song, R.; An, G. Electrochemical behavior and microstructural characterization of lanthanum-doped titanium-zirconium-molybdenum alloy. *J. Alloys Compd.* **2018**, *763*, 687–694. [CrossRef]
99. Hu, P.; Song, R.; Li, X.J.; Deng, J.; Chen, Z.Y.; Li, Q.W.; Yu, H.L. Influence of concentrations of chloride ions on electrochemical corrosion behavior of titanium-zirconium-molybdenum alloy. *J. Alloys Compd.* **2017**, *708*, 367–372. [CrossRef]
100. Davidson, J.; Kovacs, P. Biocompatible Low Modulus Titanium Alloy for Medical Implants. U.S. Patent 5,169,597, 8 December 1992.
101. O'Brien, B.; Stinson, J.; Carroll, W. Development of a new niobium-based alloy for vascular stent applications. *J. Mech. Behav. Biomed. Mater.* **2008**, *1*, 303–312. [CrossRef]
102. Arciniegas, M.; Manero, J.M.; Peña, J.; Gil, F.J.; Planell, J.A. Study of new multifunctional shape memory and low elastic modulus Ni-free Ti alloys. *Metall. Mater. Trans. A* **2008**, *39*, 742–751. [CrossRef]
103. Arciniegas, M.; Peña, J.; Manero, J.M.; Paniagua, J.C.; Gil, F.J. Quantum parameters for guiding the design of Ti alloys with shape memory and/or low elastic modulus. *Philos. Mag.* **2008**, *88*, 2529–2548. [CrossRef]
104. Okazaki, Y. A new Ti-15Zr-4Nb-4Ta alloy for medical applications. *Curr. Opin. Solid State Mater. Sci.* **2001**, *5*, 45–53. [CrossRef]
105. Saji, V.S.; Jeong, Y.H.; Choe, H.C. A comparative study on corrosion behavior of Ti-35Nb-5Ta-7Zr, Ti-6Al-4V and CP-Ti in 0.9 wt% NaCl. *Corros. Sci. Technol.* **2009**, *8*, 139–142.
106. Rajabi, F.; Hanzaki, A.Z.; Abedi, H.R.; Farghadany, E. Corrosion behavior of thermo-mechanically processed biomedical Ti-29Nb-13Ta-4.6Zr. *J. Alloys Compd.* **2017**, *725*, 23–31. [CrossRef]
107. Popa, M.; Vasilescu, E.; Drob, P.; Raducanu, D.; Moreno, J.M.C.; Ivanescu, S.; Vasilescu, C.; Drob, S.I. Microstructure, mechanical, and anticorrosive properties of a new Ti-20Nb-10Zr-5Ta alloy based on nontoxic and nonallergenic elements. *Met. Mater. Int.* **2012**, *18*, 639–645. [CrossRef]
108. Hussein, M.A.; Kumar, A.M.; Yilbas, B.S.; Al-Aqeeli, N. Laser nitriding of the newly developed Ti-20Nb-13Zr at.% biomaterial alloy to enhance its mechanical and corrosion properties in simulated body fluid. *J. Mater. Eng. Perform.* **2017**, *26*, 5553–5562. [CrossRef]
109. Huang, L.; Chen, T.H.; Chiang, C.C.; Lin, S.Y. Electrochemical corrosion and mechanical properties of two biomedical titanium alloys. *Int. J. Electrochem. Sci.* **2018**, *13*, 2779–2790. [CrossRef]
110. Macdonald, D. The point defect model for the passive state. *J. Electrochem. Soc.* **1992**, *139*, 3434. [CrossRef]
111. Moraes, P.E.; Contieri, R.J.; Lopes, E.S.; Robin, A.; Caram, R. Effects of Sn addition on the microstructure, mechanical properties and corrosion behavior of Ti–Nb–Sn alloys. *Mater. Charact.* **2014**, *96*, 273–281. [CrossRef]
112. Hallab, N.J.; Anderson, S.; Stafford, T.; Glant, T.; Jacobs, J.J. Lymphocyte responses in patients with total hip arthroplasty. *J. Orthop. Res.* **2005**, *23*, 384–391. [CrossRef] [PubMed]
113. González, M.; Peña, J.; Gil, F.J.; Manero, J.M. Low modulus Ti–Nb–Hf alloy for biomedical applications. *Mater. Sci. Eng. C* **2014**, *42*, 691–695. [CrossRef]
114. Jacobs, J.J.; Gilbert, J.L.; Urbani, R.M. Corrosion of metal orthopaedic implants. *J. Bone Jt. Surg.* **1989**, *80*, 268–282. [CrossRef]
115. Gittens, A.; Olivares-Navarrete, R.; Tannenbaum, R.; Boyan, B.D.; Schwartz, Z. Electrical implications of corrosion for osseointegration of titanium implants. *J. Dent. Res.* **2011**, *90*, 1389–1397. [CrossRef]
116. Bhola, R.; Bhola, S.M.; Mishra, B.; Olson, D.L. Corrosion in titanium dental implants/prostheses—A review. *Trends Biomater. Artif. Organs* **2011**, *25*, 34–46.
117. Manivasagam, G.; Dhinasekaran, D.; Rajamanickam, A. Biomedical implants: Corrosion and its prevention—A review. *Recent Patents Corros. Sci.* **2010**, *2*, 40–54. [CrossRef]
118. Sargeant, A.; Goswami, T. Hip implants: Paper V. Physiological effects. *Mater. Des.* **2006**, *27*, 287–307. [CrossRef]

MDPI
St. Alban-Anlage 66
4052 Basel
Switzerland
Tel. +41 61 683 77 34
Fax +41 61 302 89 18
www.mdpi.com

Metals Editorial Office
E-mail: metals@mdpi.com
www.mdpi.com/journal/metals

www.ingramcontent.com/pod-product-compliance
Lightning Source LLC
Chambersburg PA
CBHW041951260926
51850CB00019B/2416
9783039364992